Harvey Rishikof, Stewart

MW00462252

AMERICAN B
STANDING COMMITTEE ON LAW AND NATIONAL SECURITY

PATRIOTS
DEBATE

Contemporary Issues in National Security Law

Cover design by ABA Publishing

16 15 14   5 4 3

Library of Congress Cataloging-in-Publication Data
Patriots debate.
    p. cm.
    Includes bibliographical references and index.
    1. Executive power—United States. 2. Terrorism—United States—Prevention. 3. Data protection—Law and legislation—United States. 4. National security—Law and legislation—United States. 5. Commission on Accreditation for Law Enforcement Agencies. 6. Information warfare—United States. 7. Cyberterrorism—United States—Prevention. 8. Detention of persons—United States. 9. War on Terrorism, 2001-2009. 10. War and emergency powers—United States. I. American Bar Association.
KF5053.P38 2012
344.7305'25—dc23   2012021479

ISBN: 978-1-61438-590-5

Discounts are available for books ordered in bulk. Special consideration is given to state bars, CLE programs, and other bar-related organizations. Inquire at Book Publishing, ABA Publishing, American Bar Association, 321 North Clark Street, Chicago, Illinois 60654.

www.ShopABA.org

Table of Contents

Contents

Preface

ABA President-Elect Laurel G. Bellows

The American Bar Association has long held the tradition that lawyers have an important role to play in framing the legal debate on issues of importance to our nation. The intersection of law and national security has become one of the most contested issues in the past century as Americans struggle to protect our Constitutional values.

I commend the ABA Standing Committee on Law and National Security for assembling some of the brightest minds in national security and privacy today and presenting this series of debates on topics that will define our nation for decades to come.

As lawyers, it is important that we speak out on issues that traditionally define us as Americans and affect the rights and liberties of all. To speak out effectively, we must examine both sides of complex national security issues to fully comprehend the stakes. *Patriots Debate: Contemporary Issues in National Security Law* presents this balanced view.

The Committee's mission of achieving national security under law has never been more important. I offer warm greetings to the Standing Committee on the occasion of its 50th-year anniversary and congratulations on this important debate.

Introduction

Harvey Rishikof

I t is with great pleasure that the Standing Committee on Law and National Security (SCOLANS) presents *Patriots Debate*. This series of debates is a sequel to our earlier volume, *Patriot Debates: Experts Debate the USA PATRIOT Act*, published in 2005. The first book in this series was mainly concerned with the USA PATRIOT Act—the Uniting (and) Strengthening America (by) Providing Appropriate Tools Required (to) Intercept (and) Obstruct Terrorism Act of 2001. It was widely acclaimed for promoting serious public discussion about the more controversial sections of the Act.

The idea to reprise the format for some of today's "hot" issues in national security in the debate structure should be largely credited to Bernie Horowitz, one of our co-editors. Bernie approached SCOLANS with the idea of a new book and we settled on the following 10 issues—Executive Power; The Threat From Within: Homegrown Terrorism; Interrogations; Third-Party Information; National Security Letters; Einstein 3.0; the Communications Assistance Law Enforcement Act of 1994 (CALEA); a Legal Framework for Targeted Killing; Cyberwar; and Detention Policies.

As one can see, the issues fall into three general discrete categories. Part One, the War on Terrorism (or the struggle against violent extremism), has generated a broad assertion of executive power by two administrations over the last 10 years. The case for expansive executive power hinges in part on the definition of the threat. The debate between John Yoo and Louis Fisher focuses on what authority should be accorded to the President in exercising his war powers—the fear of the imperial presidency is contrasted with the concept of executive necessity and authority in the face of the inherent powers of congressional control of the purse.

The second debate in this section is "The Threat From Within." Gordon Lederman details the potential insider threat posed by violent Islamic extremism fueled by Internet-based radicalization and cites the recent case of Major Nidal Hasan as one of 32 cases since May 2009 that have increased the FBI's focus on domestic terrorism. Kate Martin, in her response, questions the scope of the threat and addresses which resources ought to be utilized and trade-offs made, and raises the issue of risks to ethnic or religious minorities.

Part One of *Patriots Debate* ends with an exploration of terrorism interrogations. Norman Abrams argues in favor of an exigent circumstance *New York v. Quarles*–based model for interrogations, weighing the relative "coerciveness" of the interrogation with the potential for catastrophic consequences. In response, Christopher Slobogin contends that current law already grants enough leeway for interrogating suspects; "guile and graft" are sufficient, he says, and such an exception would make for bad policy given the tendency for "mission creep" issues.

Part Two of *Patriots Debate* focuses on data, technology, and privacy. Increasingly in the modern age, surveillance and data processing have become the *sine qua non* for the prevention of terrorism. Concomitantly, new technology has challenged old laws. In this section we explore how government has responded to these challenges in four different but related exchanges. The Third-Party Information debate between Greg Nojeim and Orin Kerr raises critical Fourth Amendment questions in a new age where seas of information held by third parties are left unresolved in *United States v. Jones*. What constitutes a reasonable expectation of privacy as technology expands its power to penetrate physical objects? How do we distinguish between "non-content" and "content" information? When third parties are mere conduits for information, does the individual still retain a right of privacy?

Introduction

The privacy theme is continued in the controversy over National Security Letters (NSLs) generated by their expansion with the changed standards for issuance of "relevance" by the original PATRIOT Act under Section 505. The debate is conducted on one side by Michael German and Michelle Richardson, with Valerie Caproni and Steven Siegel on the other. Citing Department of Justice IG reports, German and Richardson catalogue the abuses and misuses of the NSLs by the FBI. Caproni and Siegel stress the limited nature of NSLs, the reforms made in the wake of the IG reports, the critical nature of the Attorney General's Guidelines in national security investigations, the types of information gleaned from third parties, and the new "gag" rules in the wake of recent case law.

Rounding out Part Two of *Patriots Debate,* chapters are devoted to debates examining Einstein 3.0 and the Communications Assistance Law Enforcement Act (CALEA). The active protection of the federal cyber infrastructure forms the heart of the Einstein 3.0 discussion between Paul Rosenzweig and James Dempsey, a debate over who should monitor privately owned and operated networks for cybersecurity risks.

Does the federal government have the capabilities and agility to undertake these functions? If so, would that be appropriate? Is the Defense Industrial Base the harbinger of the future? Finally, in this section, Tony Rutkowski and Susan Landau debate the future of CALEA and explore the issue of determining electronic telecommunications surveillance standards in the evolving world of telecommunications technology and the problem of "going dark." How the FCC can technically and legally solve legitimate law enforcement requirements without increasing cybersecurity vulnerabilities is part of the riddle in this vibrant discussion.

Part Three of *Patriots Debate* explores the legal frameworks for how America projects force and how we will go forward with the detainment issue. We begin this section with the Targeted Killing debate between Amos Guiora and Monica Hakimi, who examine the criteria and standards informing the elusive legal concepts of "imminence" and "self-defense," which constitute the heart of the targeting issue. This exchange is followed by Stewart Baker and General Charles Dunlap, USAF, ret., who debate how to reconcile operations in cyberspace with the laws of armed conflict. The advantages, disadvantages, and consequences of a "no-legal-limits" approach are explored and reviewed. We conclude this section with the problem that has been with us since the first prisoner was captured. Stephen Vladeck and Greg Jacob debate the effects of *Boumediene v. Bush* and *al-Maqaleh v. Gates* on targeted killing, indefinite detention, and whether the courts have struck the right balance for the standard of judicial review of combatants.

This project was supported by many people. In particular, we would like to thank Stewart Baker, editor of the first *Patriot Debates*, for his assistance in pulling this together. We also wish to express appreciation to the authors, some of the best in the field, for their time and commitment to a passionate and civil debate. We would also like to thank the *ABA Journal* for serializing a number of the debates in the magazine. Over the last 50 years, it has been and continues to be the goal of SCOLANS to foster expert-led public discussions on hard issues to improve the understanding of the principles at stake. We hope *Patriots Debate* upholds this tradition. As part of our goal to continue to stimulate debate on these subjects, we have dedicated a section of the committee website— www.americanbar.org/natsecurity—to give readers the opportunity to participate in the debates by sending comments to patriotdebates@gmail.com. Some of these comments will be displayed on the book's companion website. We encourage you to send your thoughts and responses to this moderated website so that in the end the best policies will prevail.

Part One:
The War on Terrorism

Chapter One
Executive Power

John Yoo
Louis Fisher

Introduction

C ommanding the American Warship the USS *Constitution* in
the waning days of the 18th century, Captain Silas Talbot
steered the massive 44-gun frigate into the path of a little
vessel called the *Amelia*.

Talbot knew what he was after. An experienced naval officer
since the American Revolution whose prowess prompted President
George Washington to appoint him to supervise the *Constitution*'s
construction, Talbot sought to put a potentially dangerous ship out of
commission.

But he also wanted what Congress seems to have promised: a fee
from the salvage of the *Amelia*. Congress had passed statutes that
appeared to have allowed for the seizing of such ships during the
Quasi-War, a naval confrontation in the 1790s between the newly
formed United States and France, its former ally in the American
Revolution. France and England were openly warring with each other,
and Americans were arguing over whether to stay neutral.

The *Amelia* was sailing from Calcutta, an English possession,
with a cargo of produce and dry goods. The ship was owned by a
mercantile company in the city-state of Hamburg, which was neutral
in the English-French war.

A French corvette seized the *Amelia* in September of 1799, add-
ing a crew and arms, and sailed it to the West Indies as a naval prize.
A week later, the *Constitution* caught up with the *Amelia* and captured it. The Hamburg company's representative, Hans Frederic

Seeman, wanted the ship back; Talbot claimed it was a legitimate prize of the conflict with France and sued in federal court.

The case, *Talbot v. Seeman,* which came before the U.S. Supreme Court in 1801, was remarkable in several ways. First, it revealed the growing split between America's emerging political parties. The case was argued in the appellate court by two archenemies: Federalist Alexander Hamilton, whose party sought war with France, for Talbot; and for Seeman, Aaron Burr, a Republican whose party wanted to avoid foreign alliances and defend neutral shipping.

Second, it was among the first cases in which new Chief Justice John Marshall employed a novel method of decision making. He changed the age-old routine of a series of justices' opinions into a single opinion of the Court.

And, most important, the *Talbot* case laid part of the groundwork for what has become one of the most profound debates of American war policy: whether the Constitution requires war to be declared by Congress or whether the President has the sole authority to engage in war, followed only by Congress's latent approval via the power of the purse.

Here, the confrontation moves from the floating *Constitution* to the written one. It also sets the stage for a discussion between two prominent constitutional law scholars, Louis Fisher and John Yoo.

— Richard Brust
Staff writer, *A.B.A. Journal*

Constitutional War and the Political Process

John Yoo

In ordering the U.S. air force to attack Libyan targets on the ground and imposing a no-fly zone in the air, President Barack Obama has now sent the U.S. military into combat without Congress's blessing. This was not always President Obama's view. Anti-war Democrats vigorously challenged President George W. Bush's conduct of the wars in Afghanistan and Iraq by claiming that he had violated Congress's right to declare war. As a presidential candidate in 2007,

Obama once agreed: "The President does not have power under the Constitution to unilaterally authorize a military attack in a situation that does not involve stopping an actual or imminent threat to the nation."

Fast-forward four years. In announcing the intervention in Libya, President Obama told Congress that he was acting "pursuant to my constitutional authority to conduct U.S. foreign relations and as Commander in Chief and Chief Executive." As the Libyan war reached its 60th day at the end of May of 2011, President Obama sent a letter to Congress that reported on progress but did not seek any authorization.

This time, President Obama has the Constitution about right. His exercise of war powers rests firmly in the tradition of American foreign policy. Throughout our history, neither presidents nor Congresses have acted under the belief that the Constitution requires a declaration of war before the United States can conduct military hostilities abroad. We have used force abroad more than 100 times but declared war in only five cases: the War of 1812, the Mexican-American and Spanish-American Wars, and World Wars I and II.

Without any congressional approval, presidents have sent forces to battle Indians, Barbary pirates, and Russian revolutionaries; to fight North Korean and Chinese Communists in Korea; to engineer regime changes in South and Central America; and to prevent human rights disasters in the Balkans. Other conflicts, such as the 1991 Persian Gulf War, the 2001 invasion of Afghanistan, and the 2003 Iraq War, received legislative "authorization" but not declarations of war. The practice of presidential initiative, followed by congressional acquiescence, has spanned both Democratic and Republican administrations and reaches back from President Obama to Presidents Abraham Lincoln, Thomas Jefferson, and George Washington.

Common sense does not support replacing the way our Constitution has worked in wartime with a radically different system that mimics the peacetime balance of powers between the President and Congress. If the issue were the environment or Social Security, Congress would enact policy first and the President would faithfully implement it second. But the Constitution does not duplicate this system in war. Instead, our Framers decided that the President would play the leading role in matters of national security.

Those in the pro-Congress camp call on the anti-monarchical origins of the American Revolution for support. If the Framers rebelled against King George III's dictatorial powers, surely they would not give the President much authority. It is true that the revolutionaries rejected the royal prerogative, and they created weak executives at the state level. Americans have long turned a skeptical eye toward the growth of federal powers. But this may mislead some to resist the fundamental difference in the Constitution's treatment of domestic and foreign affairs. For when the Framers wrote the Constitution in 1787, they rejected these failed experiments and restored an independent, unified chief executive with its own powers in national security and foreign affairs.

The most important of the President's powers are as commander-in-chief and chief executive. As Alexander Hamilton wrote in *Federalist 74,* "The direction of war implies the direction of the common strength, and the power of directing and employing the common strength forms a usual and essential part in the definition of the executive authority." Presidents should conduct war, he wrote, because they could act with "decision, activity, secrecy, and dispatch." In perhaps his most famous words, Hamilton wrote: "Energy in the executive is a leading character in the definition of good government. . . . It is essential to the protection of the community against foreign attacks."

The Framers realized the obvious. Foreign affairs are unpredictable and involve the highest of stakes, making them unsuitable to regulation by preexisting legislation. Instead, they can demand swift, decisive action, sometimes under pressured or even emergency circumstances, that are best carried out by a branch of government that does not suffer from multiple vetoes or is delayed by disagreements. Congress is too large and unwieldy to take the swift and decisive action required in wartime. Our Framers replaced the Articles of Confederation, which had failed in the management of foreign relations because it had no single executive, with the Constitution's single president for precisely this reason. Even when it has access to the same intelligence as the executive branch, Congress's loose, decentralized structure would paralyze American policy while foreign threats grow.

Congress has no political incentive to mount and see through its own wartime policy. Members of Congress, who are interested in keeping their seats at the next election, do not want to take stands on controversial issues where the future is uncertain. They will avoid like the plague any vote that will anger large segments of the electorate. They prefer that the President take the political risks and be held accountable for failure.

Congress's track record when it has opposed presidential leadership has not been a happy one. Perhaps the most telling example was the Senate's rejection of the Treaty of Versailles at the end of World War I. Congress's isolationist urge kept the United States out of Europe at a time when democracies fell and fascism grew in their place. Even as Europe and Asia plunged into war, Congress passed the Neutrality Acts designed to keep the United States out of the conflict. President Franklin Roosevelt violated those laws to help the Allies and draw the nation into war against the Axis powers. While pro-Congress critics worry about a president's foreign adventurism, the real threat to our national security may come from inaction and isolationism.

Many point to the Vietnam War as an example of the faults of the "imperial presidency." Vietnam, however, could not have continued without the consistent support of Congress in raising a large military and paying for hostilities. And Vietnam ushered in a period of congressional dominance that witnessed American setbacks in the Cold War, and the passage of the ineffectual War Powers Resolution. Congress passed the Resolution in 1973 over President Richard Nixon's veto, and no president, Republican or Democrat, George W. Bush or Barack Obama, has ever accepted the constitutionality of its 60-day limit on the use of troops abroad. No federal court has ever upheld the resolution. Even Congress has never enforced it.

Despite the record of practice and the Constitution's institutional design, critics nevertheless argue for a radical remaking of the American way of war. They typically base their claim on Article I, Section 8, of the Constitution, which gives Congress the power to "declare War." But these observers read the 18th-century constitutional text through a modern lens by interpreting "declare War" to mean "start war." When the Constitution was written, however, a declaration of war served diplomatic notice about a change in legal relations between nations. It had little to do with launching hostilities. In the century before the Constitution, for example, Great Britain—where the Framers got the idea of the declare-war power—fought numerous major conflicts but declared war only once beforehand.

Our Constitution sets out specific procedures for passing laws, appointing officers, and making treaties. There are none for waging war, because the Framers expected the President and Congress to struggle over war through the national political process. In fact, other parts of the Constitution, properly read, support this reading. Article I, Section 10, for example, declares that the states shall not "engage" in war "without the consent of Congress" unless "actually invaded, or in such imminent Danger as will not admit of delay." This provision creates exactly the limits desired by anti-war critics, complete with an exception for self-defense. If the Framers had wanted to require congressional permission before the President could wage war, they simply could have repeated this provision and applied it to the executive.

Presidents, of course, do not have complete freedom to take the nation to war. Congress has ample powers to control presidential policy, if it wants to. Only Congress can raise the military, which gives it the power to block, delay, or modify war plans. Before 1945, for example, the United States had such a small peacetime military that presidents who started a war would have to go hat in hand to Congress to build an army to fight it. Since World War II, it has been Congress that has authorized and funded our large standing military, one primarily designed to conduct offensive, not defensive, operations (as we learned all too tragically on

9/11) and to swiftly project power worldwide. If Congress wanted to discourage presidential initiative in war, it could build a smaller, less offensive-minded military.

Congress's check on the presidency lies not just in the long-term raising of the military. It also can block any immediate armed conflict through the power of the purse. If Congress feels it has been misled in authorizing war, or it disagrees with the President's decisions, all it need do is cut off funds, either all at once or gradually. It can reduce the size of the military, shrink or eliminate units, or freeze supplies. Using the power of the purse does not even require affirmative congressional action. Congress can just sit on its hands and refuse to pass a law funding the latest presidential adventure, and the war will end quickly. Even the Kosovo War, which lasted little more than two months and involved no ground troops, required special funding legislation.

The Framers expected Congress's power of the purse to serve as the primary check on presidential war. During the 1788 Virginia ratifying convention, Patrick Henry attacked the Constitution for failing to limit executive militarism. James Madison responded: "The sword is in the hands of the British king; the purse is in the hands of the Parliament. It is so in America, as far as any analogy can exist." Congress ended America's involvement in Vietnam by cutting off all funds for the war.

Our Constitution has succeeded because it favors swift presidential action in war, later checked by Congress's funding power. If a president continues to wage war without congressional authorization, as in Libya, Kosovo, or Korea, it is only because Congress has chosen not to exercise its easy check. We should not confuse a desire to escape political responsibility for a defect in the Constitution.

A radical change in the system for making war might appease critics of presidential power. But it could also seriously threaten American national security. To forestall another 9/11 attack, or to take advantage of a window of opportunity to strike terrorists or rogue nations, the executive branch needs flexibility. It is not hard to think of situations where congressional consent cannot be obtained in time to act. Time for congressional deliberation, which leads only to passivity and isolation and not smarter decisions, will come at the price of speed and secrecy.

The Constitution creates a presidency that can respond forcefully to prevent serious threats to our national security. Presidents can take the initiative and Congress can use its funding power to check them. Instead of demanding a legalistic process to begin war, the Framers left war to politics. As we confront the new challenges of terrorism, rogue nations, and WMD proliferation, now is not the time to introduce sweeping, untested changes in the way we make war.

Unconstitutional Presidential Wars

Louis Fisher

John Yoo and I agree that the Framers rejected the British system of royal prerogative. William Blackstone in his *Commentaries* placed all foreign policy and the war power with the Executive. Clearly the Framers repudiated that model, which is obvious simply by looking at the text of the U.S. Constitution. Not a single one of Blackstone's prerogatives—declaring war, making treaties, issuing letters of marque and reprisal, appointing ambassadors, raising and regulating fleets and armies, etc.—is vested in the President. They are either given expressly to Congress in Article I or are shared between the President and the Senate (treaties and appointments).

Where John Yoo and I differ is the scope accorded to the President over the war power. Yoo says that "our Framers decided that the President would play the leading role in matters of national security" and that the current war in Libya, initiated by President Obama, "rests firmly in the tradition of American foreign policy." The record shows, however, that all major wars from 1789 to 1950 were either authorized or declared by Congress. No president during that period believed that he could unilaterally take the country from a state of peace to a state of war. It was necessary to come to Congress to seek prior approval.

Federal courts understood that principle as well. In *Talbot v. Seeman* (1801), Chief Justice John Marshall wrote: "The whole powers of war being, by the constitution of the United States, vested in Congress, the acts of that body can alone be resorted to as our guides in this inquiry." It might be argued that when the delegates at the Philadelphia Convention changed the constitutional text from to "make war" to "declare war," they limited Congress to declaring war and allowed the President to make war. That was never the understanding. The Framers simply acknowledged that the President needed to "repel sudden attacks" without waiting for prior congressional authority, especially when Congress was not in session. That was a *defensive*, not an offensive, power. The latter judgment remained with Congress.

The fact that Congress retained authority to both declare and make war (i.e., *initiate* war) is clearly expressed in court rulings. A circuit court in *United States v. Smith* (1806) rejected the idea that a President or his assistants could unilaterally authorize military adventures against foreign governments. The court put the matter bluntly: "Does [the President] possess the power of making war? That power is exclusively vested in Congress." If a nation invaded the United States, the President would have an obligation to resist with force. But there was a "manifest distinction" between going to war with a nation at peace and responding to an actual invasion: "In the former case, it is the exclusive province of Congress to change a state of peace into a state of war."

It is frequently argued that the Supreme Court in *The Prize Cases* (1863) recognized a broad war power for the President. It did not. As with the *Smith* case, Justice Robert C. Grier carefully limited the President's power to defensive actions, in this case a civil war. The President "has no power to initiate or declare a war against either a foreign nation or a domestic State." During oral argument, the attorney for the administration, Richard Henry Dana, Jr., agreed that the actions of President Lincoln had nothing to do with "the right *to initiate a war, as a voluntary act of sovereignty*. That is vested only in Congress."

This understanding prevailed for 160 years, from 1789 to 1950. As Yoo notes, presidents during that period used "force abroad more than 100 times" without a declaration or authorization from Congress. But those actions, however noteworthy, did not constitute major wars. What happened in 1950 to change this constitutional pattern? It was President Truman unilaterally taking the country to war against North Korea. At no time did he come to Congress, as with all presidents in the past, to seek a declaration or authorization. If Congress did not authorize this war, who did? President Truman claimed that two resolutions passed by the U.N. Security Council provided sufficient authority.

Here I think Yoo and I would agree on a vital point. It is constitutionally impermissible for the President and the Senate through the treaty process to take power from future Senates and from the House of Representatives and give it to an outside body. In other words, the President and the Senate, in agreeing to the U.N. Charter, could not amend the Constitution by placing with the Security Council war powers that had rested with Congress. That is essentially the argument presented by President Truman and his Secretary of State, Dean Acheson. It is a shallow and empty argument. Here I would not hesitate to fault Congress for accepting that rationale and failing to protect its institutional powers. The Framers expected each branch to fight off encroachments. Congress decided it was more important to fight communism than to defend the Constitution.

There were no reasons to do so. The fundamental duty of lawmakers, as reflected in their oath of office, is to support and defend the Constitution at all times. They take that obligation "freely, without any mental reservation or purpose of evasion." There was no justification violating the Constitution because fighting North Korea had a higher value. For more than a century and a half, lawmakers had balanced the Constitution and war powers without sacrificing one for the other.

Just as President Truman had no authority for substituting the Security Council for Congress, so did President Obama violate the Constitution by arguing that he could obtain "authorization" from a

Security Council resolution to use military force against Libya. President Obama often says it is his duty to defend the country. His first duty—as reflected in the oath of office placed in Article II—is to "preserve, protect and defend the Constitution of the United States." The duty to defend the country is the duty to repel attacks on the country and its forces. Libya did not attack or threaten the United States. President Obama acted not in a defensive manner but offensively against another country. Any president who takes the country from a state of peace to a state of war without obtaining prior authority from Congress is creating an impeachable act. Were the President impeached by the House and removed by the Senate, the signal would be healthy and welcome for constitutional government.

Yoo points to some qualities that the President possesses and Congress does not. He recalls what Alexander Hamilton said about unity, decisiveness, secrecy, and energy residing in the President, not in the legislative branch. Generally, this is true. Yoo says that Congress "is too large and unwieldy to take the swift and decisive action required in wartime." The institutional advantage here clearly rests with the President. But the Framers did not give their blessing to presidential decisiveness and speed of action. They knew all too well the record of executive wars in the past, which devastated nations and left them poorer. John Jay in *Federalist 4* warned that "nations in general will make war whenever they have a prospect of getting any thing by it; nay absolute monarchs will often make war when their nations are to get nothing by it," but for purposes merely personal, such as an executive's thirst for military glory, revenge for personal affronts, and ambition. Those and other motivations lead executives "to engage in wars not sanctified by justice or the voice and interests of his people." Because of those costs, the Framers insisted that going to war be done by legislative deliberation and support.

Did the Framers have a narrow, 18th-century vision that has no application to America and contemporary conditions? It would be hard to make that argument after the presidential wars that followed World War II, including Korea, Vietnam, and the Iraq War that began in March of 2003. The Korean War was a limited effort to protect the division between the North and the South that changed once U.S. forces decided to go into North Korea and provoke the Chinese to enter. The result was a war that did substantial damage to the United States and certainly to the presidency of Harry Truman. Vietnam was a great calamity, fueled in large part by the claim that a "second attack" occurred in the Gulf of Tonkin to justify U.S. retaliation. There were doubts at the time that the second attack happened. We now know that there was no second attack, but merely late signals coming from the first. The decision to go to war against Iraq in 2003 rested on a number of claims that proved false: aluminum tubes used for making nuclear weapons, "yellowcake" obtained from a nation in North Africa, mobile labs capable of carrying biological agents, drones able to deliver chemical and biological agents, etc. In every case, those assertions by the Bush administration were vacuous. The record of presidents using deception and stealth to go to war would not surprise the Framers. They should not surprise us.

Yoo states that "Congress's track record when it has opposed presidential leadership has not been a happy one." There is some truth to that. I would not, however, fault the Senate for rejecting the Treaty of Versailles. The problem was the rigidity of President Woodrow Wilson in opposing Senator Henry Cabot Lodge's reservations. President Wilson's advisers urged acceptance of the reservations but he preferred to let personal animosities prevail over the treaty. Yoo is on stronger ground in criticizing the neutrality policy of Congress during the 1930s when fascism was sweeping Europe. Congress does not have a happy track record, but neither do many of our presidents. I would also say that Congress has done great damage to the nation and constitutional government by acquiescing to presidential wars on the mistaken belief that presidents invariably act in the "national interest" and are surrounded by officials with reliable expertise and judgment. That is a fanciful view not supported by the record. Finally, Yoo points out that Congress possesses the power of the purse to stop wars begun by presidents. I think we all know that the spending tool is difficult to invoke when U.S. soldiers are in combat. It took nearly a decade to cut off funds for the Vietnam War.

Chapter Two
The Threat from Within: Homegrown Terrorism

Gordon Lederman

Kate Martin

Introduction

This Committee cannot live in denial—which is what some would have us do when they suggest that this hearing dilute its focus by investigating threats unrelated to Al Qaeda. . . . There is no equivalency of threat between al Qaeda and neo-Nazis, environmental extremists or other isolated madmen. Only al Qaeda and its Islamist affiliates in this country are part of an international threat to our nation. Indeed by the Justice Department's own record not one terror related case in the last two years involved neo-Nazis, environmental extremists, militias or anti-war groups . . . there are realities we cannot ignore . . . a Pew Poll said that 15% of Muslim-American men between the age of 18 and 29 could support suicide bombings. This is the segment of the community al Qaeda is attempting to recruit. . . . These include New York City Subway bomber Najibullah Zazi; Fort Hood terrorist U.S. Army Major Nidal Hasan; Colleen LaRose, known as "Jihad Jane"; Times Square bomber Faisal Shahzad; Mumbai plotter David Headley; Little Rock Recruiting Center shooter Carlos Bledsoe; and dozens of individuals in Minneapolis associated with the Somali terrorist organization al Shabaab.[1]

> —*Statement of Representative Peter T. King (R-NY),*
> *March 10, 2011*

. . . demanding a "community response" (as the title of this hearing suggests) asserts that the entire community bears responsibility for the violent acts of individuals. Targeting the Muslim American community for the actions of a few is unjust. Actually all of us—all communities—are responsible for combating violent extremism. Singling out one community focuses our analysis in the wrong direction.[2]

> —*Statement of Representative Keith Ellison (D-Minn.),*
> *March 10, 2011*

Insider Threats:
Homegrown Terrorism in the 21st Century

Gordon Lederman

Homegrown terrorism is a major national security threat to the United States in the 21st century because of the combination of violent extremist ideology—especially violent Islamist extremism and its suicidal terrorism—with the Internet accelerating an individual's radicalization and with the proliferation of technology enabling an individual to cause mass casualties or widespread disruption. In other words, the scope of the insider threat to

---

1. Chairman Peter T. King, *The Extent of Radicalization in the American Muslim Community and That Community's Response*, Committee on Homeland Security, House of Representatives, March 10, 2011 (http://homeland.house.gov/sites/homeland.house.gov/files/03-10-11%20Final%20King%20Opening%20Statement_0.pdf).

2. Rep. Keith Ellison, *The Extent of Radicalization in the American Muslim Community and That Community's Response, Committee on Homeland Security, House of Representatives*, March 10, 2011 (http://ellison.house.gov/index.php?option=com_content&view=article&id=587:congressman-ellisons-testimony-to-the-house-committee-on-homeland-security-as-prepared-for-delivery&catid=36:keiths-blog&Itemid=44).

the United States in the 21st century is virtually unprecedented. An individual within the United States, segregated within his or her own Internet community of sympathizers, can potentially radicalize to violent extremism and then utilize modern technology to kill millions of people or hobble critical infrastructure. This insider threat requires a greater emphasis on preventing potentially catastrophic attacks before they occur.

The attacks of September 11, 2001, heightened America's consciousness to the threat posed by violent Islamist extremism. "Violent Islamist extremism" refers to the ideology advocating creation of a global state that would impose the most radical version of Islamic law and the use of violence against non-Muslim military personnel and civilians and even against Muslim opponents of this ideology. To recruit adherents, violent Islamist extremism utilizes a narrative that the West, led by the United States, is at war with Islam. The process by which individuals adopt violent Islamist extremism is commonly called *radicalization*.

Violent Islamist extremists demonstrated on 9/11 and in subsequent plots that they seek to kill large numbers of U.S. civilians domestically, and al Qaeda leadership has sanctioned the use and pursued the development of weapons of mass destruction. Moreover, the 9/11 attacks showed violent Islamist extremists' willingness to commit suicide while attacking, thus making them difficult to deter. These characteristics make violent Islamist extremism the most dangerous terrorist threat to the United States. Indeed, the Obama Administration has identified this threat, which it calls "violent extremism and terrorism inspired by al Qa'ida and its affiliates and adherents,"[3] as the "preemi-

---

3. The Obama Administration prefers "violent extremism and terrorism inspired by al Qa'ida and its affiliates and adherents" by arguing that use of the term "violent Islamist extremism" risks validating to Muslims globally that the ideology is justified by Islam and that the United States is at war with Islam. Letter from John Brennan, Deputy National Security Advisor and Assistant to the President for Homeland Security and Counterterrorism, to Senator Joseph Lieberman, April 22, 2010, at 3. That position has been criticized as ignoring that the ideology predates al Qa'ida, that there are violent Islamist extremists such as the Pakistani group Tehrik-i-Taliban (which sponsored one attack within the United States) that are not affiliates or adherents to

nent" violent extremist threat to the United States.[4]

The 9/11 hijackers came from outside the United States to attack. During the initial years after 9/11, the U.S. government assumed that the United States' experience as a melting pot of immigrants pursuing the American dream was a bulwark against U.S. citizens—native and immigrant—radicalizing to violent Islamist extremism. This assumption seemed generally correct because there were only 21 cases of homegrown violent Islamist extremism—that is, terrorist attacks or plots by U.S. citizens, permanent residents, or visitors radicalized largely within the United States—from 9/11 until April of 2009.[5]

But since May of 2009, there have been 32 cases of homegrown violent Islamist extremism.[6] Two of these post–2009 cases have resulted in deaths on U.S. soil. On June 1, 2009, Carlos Bledsoe killed one service member and wounded a second at a military recruiting station in Little Rock, Arkansas. And on November 5, 2009, 12 service members and one Department of Defense civilian were killed and 32 wounded at Fort Hood, Texas. Army-trained psychologist

---

al Qa'ida, and that the lack of specificity generates confusion within the government bureaucracy while not affecting how U.S. counterterrorism actions are perceived globally. Letter from Senator Joseph Lieberman to John Brennan, June 10, 2010. This ideology has also been called "jihadism," the term used by the Quilliam Foundation, a British think tank composed of former violent Islamist extremists that aims to counter this ideology. *See* http://www.quilliamfoundation.org/faqs.html (defining "jihadism" as "the use of violence to bring about Islamism; it is a framework for interpreting and justifying political violence around the world. Instead of understanding any given conflict as a product of local and regional contexts (social, political, economic etc.), jihadism interprets all conflicts involving Muslims through the lens of a narrative that perceives Islam as a religion to be under attack, and therefore in need of a violent defense).

4. White House, Strategic Implementation Plan for Empowering Local Partners to Prevent Violent Extremism in the United States, December 2011, at 2.

5. Jerome Bjelopera, American Jihadist Terrorism: Combating a Complex Threat (Cong. Research Service, 2011), at 6.

6. *Id.* More than half of these cases involved attacks or plots against military personnel domestically. This statistic suggests the deliberate targeting of military personnel, perhaps under the misguided belief by some American violent Islamist extremists that attacks against U.S. military personnel are more legitimate and justifiable than attacks against fellow American civilians.

Major Nidal Hasan, who reportedly had radicalized to violent Islamist extremism during his military medical training, is being court-martialed for the attack.[7] Other notable plots have included a 2008 plot led by Najibullah Zazi to attack the New York City subway system and the attempt on May 1, 2010, by Faisal Shahzad to detonate a car bomb in Manhattan's Times Square. The vast majority of these cases have not involved plans for suicide, but Zazi and his fellow plotters did plan their attack as a suicide mission, and several Americans who joined the al-Shabaab terrorist group in Somalia reportedly became suicide attackers there.[8]

These cases of homegrown violent Islamist extremism represent a tiny percentage of the estimated six million Muslim-Americans. Moreover, as Presidents George W. Bush and Barack Obama have stressed since 9/11, the United States is not at war with Islam but rather with adherents to an ideology that perverts it. However, this upward trend in cases is worrisome, particularly when combined with the role of the Internet and the proliferation of destructive and disruptive technology.[9]

The Internet has facilitated radicalization to violent Islamist extremism and resulting terrorist activity. To be sure, the Internet has provided billions of people with access to information, and social media has played a critical role in democratic revolutions across the

---

7.   U.S. SENATE COMMITTEE ON HOMELAND SECURITY AND GOVERNMENTAL AFFAIRS, A TICKING TIME BOMB: COUNTERTERRORISM LESSONS FROM THE U.S. GOVERNMENT'S FAILURE TO PREVENT THE FORT HOOD ATTACK (February 2011), at 27–34 [hereinafter A TICKING TIME BOMB].

8.   AMERICAN JIHADIST TERRORISM, *supra* note 5, at 33.

9.   Although the violent Islamist extremist threat is preeminent, it is not the only violent extremism that constitutes an insider threat. For example, the bombing of the Alfred O. Murrah Federal Building in Oklahoma City, Oklahoma, on April 19, 1995, by Timothy McVeigh showed the threat from right-wing extremists. The right-wing violent extremist threat was also demonstrated outside of the United States by the July 22, 2011, car bomb attack against the office of the Norwegian Prime Minister, combined with the killing of 69 people at a summer camp, allegedly perpetrated by a Norwegian who opposed Islam and immigration. The accelerating and self-segregating nature of the Internet, matched with the proliferation of destructive and disruptive technology, make violent right-wing extremism and other such extremisms more dangerous as well.

world. However, the Internet has also enabled radicalization, with violent Islamist extremists becoming adept at using the Internet to spread their propaganda. Violent Islamist extremists originally used password-protected forums, but they are now present on mainstream sites such as YouTube. The Internet enables individuals who are vulnerable to radicalization to find violent Islamist extremist material easily and to self-segregate and interact only with individuals who share that ideology—and in the privacy of their own homes.[10]

Homegrown violent Islamist extremist plots to date have not involved weapons of mass destruction but rather conventional explosives and firearms. Of course, firearms can cause significant casualties, such as the terrorist attack that killed 168 people in Mumbai, India, in November of 2008. But the proliferation of mass destructive and disruptive technology enables even a single individual to wreak havoc domestically. For example, a single terrorist with microbiological training and access to a laboratory with dangerous pathogens—or equipment enabling the synthesis of a new pathogen—could release a pathogen that could kill millions.[11] In addition, an individual with cyber skills could cause extensive damage to infrastructure and even loss of life through a cyber attack, such as by interfering with supervisory control and data acquisition systems that control the underpinnings of modern society, such as electrical power transmission, communications, and airports. An individual could also

10. As the administration's *Strategic Implementation Plan* states, "The Internet has become an increasingly potent element in radicalization to violence, enabling violent extremists abroad to directly communicate to target audiences in the United States. This direct communication allows violent extremists to bypass parents and community leaders." STRATEGIC IMPLEMENTATION PLAN, at 20.

11. COMMISSION ON THE PREVENTION OF WEAPONS OF MASS DESTRUCTION PROLIFERATION AND TERRORISM, WORLD AT RISK (December 2008), at 27 (stating that "the more that sophisticated capabilities, including genetic engineering and gene synthesis, spread around the globe, the greater the potential that terrorists will use them to develop biological weapons"). To illustrate the concern, in December 2011 the federal National Science Advisory Board for Biosecurity asked two scientific journals not to publish the details of an experiment that made the bird flu—already highly lethal to humans—highly contagious out of fear that terrorists would turn it into a weapon. David Brown, *Federal Panel Asks Journals to Censor Reports of Lab-Created 'Bird Flu,'* WASH. POST (Dec. 20, 2011).

gain access to radiological material—such as held by hospitals for medical purposes—and create a "dirty bomb" that renders several city blocks or larger areas uninhabitable and sow panic. The vector of technological development thus continually increases the power of an individual to cause mass destruction and disruption.

The result of the combination of violent extremism, especially violent Islamist extremism, the role of the Internet, and the proliferation of destructive technology is that the government must increasingly emphasize prevention of homegrown terrorism. This task is complicated by the fact that, although an overall four-stage model of radicalization to violent Islamist extremism does exist,[12] individuals who have radicalized have not necessarily followed the model's sequence of stages,[13] and an analysis of homegrown terrorism cases does not reveal a profile to predict who will radicalize except that homegrown terrorists are predominantly male and approximately two-thirds are under 30 years old.[14]

Immediately after the 9/11 attacks, Federal Bureau of Investigation (FBI) Director Robert Mueller declared that the FBI, which had previously focused on investigating terrorism after its occurrence, now had the top priority of preventing terrorist attacks.[15] To do so, the FBI needed to reorient itself from prosecutions after an attack to an intelligence-driven effort to detect and dismantle terrorist threats prior to an attack.[16] For example, Director Mueller charged each of

---

12. MITCHELL SILBER & ARVIN BHATT, RADICALIZATION IN THE WEST: THE HOMEGROWN THREAT, City of New York Police Department, Intelligence Division (2007), at 6–8. This model sets forth a four-step process beginning with a pre-radicalization phase; then self-identification due to a crisis or other trigger causes them to explore violent Islamist extremism, then indoctrination, and finally the transition to violent activity. The FBI's model is similar. Carol Dyer, Ryan McCoy, Joel Rodriguez et al., *Countering Violent Islamic Extremism: A Community Responsibility*, in FBI LAW ENFORCEMENT BULL. (December 2007), at 6.

13. AMERICAN JIHADIST TERRORISM, *supra* note 5, at 13.

14. David Schanzer, Charles Kurzman & Ebrahim Moosa, *Anti-Terror Lessons of Muslim Americans* (Jan. 6, 2010), at 10.

15. Robert Mueller, Director, FBI, Statement before the House Judiciary Committee (May 20, 2009).

16. Robert Mueller, Director, FBI, Statement before the Senate Judiciary Committee (March 5, 2008).

the 56 FBI field offices with "domain awareness," defined as "a 360-degree understanding of all national security and criminal threats in any given city or community. It is the aggregation of intelligence, to include what we already know and what we need to know, and the development of collection plans to find the best means to answer the unknowns. With this knowledge, we can identify emerging threats, allocate resources effectively, and identify new opportunities for intelligence collection and criminal prosecution."[17] Director Mueller also declared that every counterterrorism lead would be pursued.[18, 19]

However, prior to the 9/11 attacks, the FBI was not authorized to conduct investigative activity without sufficient factual predication that a crime was being or had been committed. For the FBI to become "intelligence-driven," it had to develop a protocol for collecting intelligence information, that is, information concerning potential threats even if there was no factual predication of criminal activity. Without such a protocol, the FBI would have limited ability to track down counterterrorism leads that lacked predication of criminal activity or to gather information to be able to analyze the nature and trend of violent extremist threats above and beyond the FBI's ongoing criminal cases.

Accordingly, in 2008, Attorney General Michael Mukasey authorized and Director Mueller instituted a new FBI operational protocol permitting "assessments" when there is "no particular factual predication" that a crime has been committed and instead based on an "authorized purpose" such as "to detect, obtain information about, or prevent or protect against federal crimes or threats to the national security or to collect foreign intelligence."[20]

---

17. Robert Mueller, Director, FBI, Statement before the Senate Judiciary Committee (March 25, 2009).

18. FBI, *Counterterrorism Division Program Management*, Electronic Communication #66F-HQ-A1308701 (Dec. 25, 2002), *cited in* A TICKING TIME BOMB, at 52.

19. AMERICAN JIHADIST TERRORISM, at 33. Shirwa Ahmed conducted a suicide bombing in Somalia in 2008, and Farah Mohamed Beledi died when he attempted to detonate his suicide vest in Somalia in 2011. *Id.*

20. FBI, DOMESTIC INVESTIGATIONS OPERATIONS GUIDE (Dec. 16, 2008), at 39, *available at* http://vault.fbi.gov/FBI%20Domestic%20Investigations%20and%20Operations%20Guide%20(DIOG).

According to the protocol:

- Limited investigative tools are permitted, including reviewing publicly available information and conducting physical surveillance not otherwise requiring a warrant, as opposed to intrusive tools such as wiretaps, which require a warrant.[21]
- Assessments are prohibited "based solely on the exercise of First Amendment protected activities or on the race, ethnicity, national origin or religion of the subject."[22]
- FBI agents must use the "least intrusive" investigative mechanisms possible.[23]

As stated in Attorney General Mukasey's authorization to the FBI, "For example, assessment activities may involve proactively surfing the Internet to find publicly accessible websites and services through which recruitment by terrorist organizations and promotion of terrorist crimes is openly taking place."[24]

The use of assessments does raise the risk of abuse, such as that the FBI's guidelines for use of assessments will be ignored in practice and that the assessments will be conducted based solely on First Amendment activity or that racial and ethnic profiling will be done. However, three counterbalancing forces provide the necessary checks:

First, FBI and other law enforcement personnel need adequate training concerning the nature of violent extremism and in particular violent Islamist extremism and how it differs from the peaceful practice of Islam. Training has often focused on behavioral indicators of radicalization—such as whether an individual is isolating himself or herself from friends and family—and not the ideology of violent Islamist extremism. As the Senate investigation of the Fort Hood attack

---

21. The Attorney General's Guidelines for Domestic FBI Operation (Sept. 29, 2008), at 20, *available at* http://www.justice.gov/ag/readingroom/guidelines.pdf.

22. *Id.*

23. Domestic Investigations Operations Guide, *at 44.*

24. The Attorney General's Guidelines for Domestic FBI Operation (Sept. 29, 2008), at 17, *available at* http://www.justice.gov/ag/readingroom/guidelines.pdf.

found, "Understanding the ideology of violent Islamist extremism would assist agents in determining, in conjunction with an individual's conduct, what degree of risk an individual might present and whether to pursue further inquiry."[25] In addition, state and local law enforcement are the first lines of defense against terrorism because of their knowledge of and constant contact with local communities, and anecdotal evidence raises concerns about the quality of training that they receive. Such training by outside experts has reportedly included that "Islam is a highly violent religion" and that if an individual has "different spellings of a name . . . that's probably cause to take them in."[26] Preventing terrorism from violent Islamist extremism requires accurate training so that law enforcement agents can focus their efforts on extremists and not on Americans practicing Islam.

Second, oversight of law enforcement intelligence activities is essential. The FBI is overseen by the Department of Justice, especially the Department of Justice Inspector General, and by various congressional committees. The Department of Justice Inspector General has issued a variety of reports including a report on FBI agents cheating on tests concerning the new FBI investigative authorities discussed above,[27] FBI investigations of domestic advocacy groups (which did not find systemic abuses),[28] and errors and management failures in the use of FBI authorities to request information from communications providers.[29] Congress exercises oversight via hearings, requests for briefings, appropriation of the FBI's funding, and

---

25. A Ticking Time Bomb, *supra* note 7, at 76.

26. *How We Train our Cops to Fear Islam*, Wash. Monthly (March 2011).

27. U.S. Dep't of Justice Inspector General, Investigations of Allegations of Cheating on the FBI's Domestic Investigations Operations Guide (DIOG) Exam (September 2010).

28. U.S. Dep't of Justice Inspector General, A Review of the FBI's Investigations of Certain Domestic Advocacy Groups (September 2010).

29. U.S. Dep't of Justice Inspector General, A Review of the Federal Bureau of Investigation's Use of Exigent Letter and Other Informal Requests for Telephone Records (January 2010); A Review of the FBI's Use of National Security Letters: Assessment of Corrective Actions and Examination of NSL Usage in 2006 (March 2008); and A Review of the Federal Bureau of Investigation's Use of National Security Letters (March 2007).

launching special investigations. The criticality of preventing home-grown terrorism, combined with the new authorities granted to the FBI, require that oversight of the FBI be rigorous and sustained.

And third, the FBI's attempts to prevent terrorist activity by uti-lizing its authorities must take place alongside vigorous U.S. govern-ment outreach to local communities and in particular Muslim-Ameri-can communities. This outreach is designed to build relationships with local communities, provide them with information that they need to develop communal efforts to prevent and respond to radicalization in their midst, and elicit concerns about particular government actions or policies, including the FBI's conduct of assessments. This outreach is conducted by a range of agencies, including the Department of Home-land Security's Civil Rights and Civil Liberties office and the FBI. The Obama Administration released an overall framework in August of 2011[30] and an implementation plan in December of 2011 that listed tasks along with leading and supporting government agencies.[31] The challenge now is to ensure that this plan is implemented aggressively and with clear leadership and commensurate resources, thus providing the greatest chance of preventing terrorism especially motivated by violent Islamist extremism.

Overlooked Constitutional and Security Issues

Kate Martin

Gordon Lederman begins his paper with a statement that, if true, could not be any more alarming:

> [T]he scope of the insider threat to the United States in the 21st century is virtually unprecedented—there exists the potential for an individual within the United States, segre-gated within his or her own Internet community of sympa-thizers, can potentially radicalize to violent extremism and

---

30. WHITE HOUSE, EMPOWERING LOCAL PARTNERS TO PREVENT VIOLENT EXTREMISM IN THE UNITED STATES (August 2011).
31. STRATEGIC IMPLEMENTATION PLAN.

then utilize modern technology to kill millions of people or hobble critical infrastructure (at 1).

But do recent events make a very strong case that there are individuals within the United States who ascribe to "violent Islamist extremism" (in Lederman's terms) and pose an existential threat to the country? Or does this kind of rhetoric obscure the difficult constitutional issues faced by law enforcement in attempting to prevent acts of domestic terrorism? Does singling out this threat have negative consequences for security as well as for fundamental civic values? And what does it mean for evaluating an appropriate government response?

Scope of the Threat

The scope of the threat is beyond my expertise as a lawyer. It is to some degree unknown, although policy makers, of course, must deal with future threats, even when unknown. Unfortunately, suggesting the existence of an existential threat, whether correct or not, makes close analysis of the real likelihood of such a threat or of the costs and benefits of possible responses thereto more difficult and even unlikely.[32]

Recent testimony by the Director of National Intelligence (DNI), James Clapper, provides a more nuanced picture of the specific threats that the intelligence community is worried about, including the possibility of a nuclear-armed Iran, the ongoing war in Afghanistan, and instability in the Middle East. Director Clapper did not include the substantial likelihood that a homegrown terrorist will kill millions of people.[33] Recent successes against al Qaeda in Afghanistan, Pakistan, and Yemen would seem to make it more difficult for would-be ter-

---

32. The response to the suggestion by officials in the former administration of the existence of a mushroom cloud on the horizon is perhaps the best recent example of this phenomenon.

33. "The Intelligence Community sees the next two or three years as a critical transition phase for the terrorist threat, particularly for al-Qa'ida and like-minded groups. With Usama bin Ladin's death, the global jihadist movement lost its most iconic and inspirational leader. The new al-Qa'ida commander is less charismatic, and the death or capture of prominent al-Qa'ida figures has shrunk the group's top leadership layer."

rorists within the United States to obtain the training, financing, and logistical support that underlay the ability of the 9/11 hijackers to pull off an attack of that magnitude.

In arguing that this threat poses a unique menace, Lederman's paper relies in large measure on the so-called rise in incidents of homegrown "Islamist" extremism in 2009. But not all analysts agree that these instances support such conclusions or predictions about the magnitude of the future risk. As the preeminent independent expert in this field concluded:

"However, even with its degraded capabilities and its focus on smaller, simpler plots, al-Qa'ida remains a threat. As long as we sustain the pressure on it, we judge that core al-Qa'ida will be of largely symbolic importance to the global jihadist movement. But regional affiliates, as the ones you mentioned, and to a lesser extent, small cells and individuals, will drive the global jihad agenda. . . .

"We assess that a mass attack by foreign terrorist groups involving a chemical, biological, radiological or nuclear (CBRN) weapon in the United States is unlikely in the next year, as a result of intense counterterrorism pressure," at 2. "The [Intelligence Community] judges that lone actors abroad or in the United States—including criminals and homegrown violent extremists (HVEs) inspired by terrorist leaders or literature advocating use of CBR materials—are capable of conducting at least limited attacks in the next year, but we assess the anthrax threat to the United States by lone actors is low," at 2.

James R. Clapper, Director of National Intelligence, Jan. 31, 2012, Unclassified Statement for the Record on the Worldwide Threat Assessment of the U.S. Intelligence Community, Senate Select Intelligence Committee, http://www.dni.gov/testimonies/20120131_testimony_ata.pdf.

Likewise, when the former Deputy Secretary for Homeland Security in the Bush administration was asked about current threats, he focused not on individuals within the United States but on the methodologies whose use poses the greatest threat, which methods are much more easily available to foreign states and groups than individuals within the United States. See Testimony of Paul A. Schneider, Former Department of Homeland Security Deputy Secretary, before the U.S. House of Representatives Committee on Homeland Security, Subcommittee on Management, Investigations and Oversight, Feb. 3, 2012: "I believe the most serious dangers facing our nation today involve biological, cyber and nuclear threats" at 1. *Available at* http://homeland.house.gov/hearing/subcommitte-hearing-dhs-effectively-implementing-strategy-counter-emerging-threats.

[My] analysis suggests that homegrown jihadists pose a terrorist threat, but thus far, despite al Qaeda's intensive online
recruiting campaign, their numbers remain small, their determination limp, and their competence poor. Even the apparent uptick in arrests in 2009 and 2010 turns out upon
close examination to be the culmination of investigations of
activity in earlier years, the result of young Somali-Ameri-
cans going off to fight Ethiopian invaders, and the honing
of the Federal Bureau of Investigation's (FBI's) investigative techniques that have led to a number of stings. The
increase is real, but it is gradual.[34]

While FBI Director Robert Mueller has taken a more dire view,
there is no consensus that the experience to date of plots in fact
supports such dire predictions.[35] And many are skeptical of the dangers of the Internet; as one expert testified, experience to date "suggests a failure of al Qaeda's Internet Strategy."[36]

---

34. B. M. Jenkins, *Stray Dogs and Virtual Armies, Radicalization and
Recruitment to Jihadist Terrorism Since 9/11*, Occasional Paper, RAND Corp.
(2011), at 1.

35. It is not clear that more than one or two of the individuals arrested in the
U.S. since 9/11 were willing to commit suicide; none of them attempted to use
CBRN weapons, although some did attempt to use home-made bombs; it is not
clear that any of them had the capability to access or use CBRN—those who
succeeded used guns; and it appears that many of them were individuals with
limited capabilities. "In terms of the potential for casualties, the bulk of the suspects
in 2011 appeared to have been limited in competence." C. KURZMAN, MUSLIM AMERICAN
TERRORISM IN THE DECADE SINCE 9/11, The Triangle Center on Terrorism and
Homeland Security, Feb. 8, 2012, at 3.

36. "Despite years of online jihadist exhortation and instruction, the level of
terrorist violence in the United States during the past decade is far below the terrorist
bombing campaigns carried out by a variety of groups in the 1970s. The absence of
jihadist terrorist activity since 9/11 reflects the success of domestic intelligence
operations. It also indicates that America's Muslim community has rejected al Qaeda's
ideology. And it suggests a failure of al Qaeda's Internet Strategy." B. M. Jenkins,
Is Al Qaeda's Internet Strategy Working?, Testimony before the House Committee
on Homeland Security, Subcommittee on Counterterrorism and Intelligence, Dec.
6, 2011, at 32.

Of course, this is not to deny the existence of domestic terrorism and its lethal consequences, nor the possibility that it could result in much greater harm in the future than it has to date. But the magnitude of the threat matters in deciding what kind and level of resources should be used, what trade-offs should be made, and how government should be structured.

Constitutional Questions

Because terrorism, unlike other violent crimes, is by definition tied to political, ideological, or religious beliefs, it is difficult, but essential, for law enforcement and intelligence rules and practices to distinguish between constitutionally protected beliefs and advocacy and criminal plans or activity. Too often, law enforcement has yielded to the temptation of focusing on those whose beliefs are outside the mainstream, rather than on those plotting terrorist acts, who are likely to be more difficult to identify.[37] The potential confusion between constitutional speech and beliefs and criminal acts is even reflected in the language describing the problem, e.g., *violent extremism*. The term is imprecise and could encompass extreme beliefs, such as belief in the necessity of and moral justification for violence, which are protected by the First Amendment and constitutionally distinct from violent acts, which are not protected.[38] The term *extremist violence* would be more precise and less likely to lead to constitutional confusion by government officials.

In recent years, there have been many reports of government bias against Islam or Muslims, including the use of false and bigoted

---

37. Just last month, the press reported law enforcement identifying protected speech and beliefs as identifiers of potential terrorists, focusing on the fact that terrorists "increasingly spoke out against the government, blamed the government for perceived problems and did so in a way that caught the attention of other people in their communities" E. Sullivan, *Obama Administration Holding Terrorism Summit with Police Chiefs*, THE HUFFINGTON POST, Jan. 18, 2012, *available at* http://www.huffingtonpost.com/2012/01/18/obama-administration-police-chiefs-violent-extremism_n_1212697.html.

38. By acts of violence, I include, of course, attempts and conspiracies to commit such violence.

law enforcement training materials[39] and widespread surveillance of mosques, which suggest at a minimum the absence of a clear understanding of constitutional limits.[40]

---

39. Michael S. Schmidt & Charlie Savage, *Language Deemed Offensive Is Removed From F.B.I. Training Materials*, N.Y. TIMES (March 28, 2012), *available at* http://www.nytimes.com/2012/03/29/us/politics/language-deemed-offensive-is-removed-from-fbi-training-materials.html.

40. The FBI's use of bigoted training materials was uncovered through media reporting and Freedom of Information Act requests. Since then, the FBI has agreed to purge the materials and is reportedly finding hundreds of pages of such documents. Spencer Ackerman, *FBI Purges Hundreds of Terrorism Documents in Islamophobia Probe*, WIRED, Feb. 12, 2012, *available at* http://www.wired.com/dangerroom/2012/02/hundreds-fbi-documents-muslims/; the ACLU's website hosts the documents obtained through FOIA: *see* http://www.aclu.org/national-security/aclu-eye-fbi-fbis-use-anti-arab-and-anti-muslim-counterterrorism-training. Senator Richard Durbin described some of the materials during a Senate Judiciary Committee Hearing: "[W]e have found that the FBI agents who were given counterterrorism training were, unfortunately, subjected to many stereotypes of Islam and Muslims. For example, FBI agents in training were told that, quote, 'Islam is a highly violent radical religion,' quote, 'mainstream American Muslims are likely to be terrorist sympathizers,' and, quote, 'the Arabic mind is swayed more by ideas than facts.'" In response, Attorney General Eric Holder condemned that training and promised: "[We] have a process under way to review the materials to make sure that that mistake does not happen in the future." Senate Judiciary Committee Hearing on Oversight of the U.S. Dep't of Justice, Nov. 8, 2011, *available at* http://www.judiciary.senate.gov/hearings/hearing.cfm?id=9b6937d5e931a0b792d258d9b32d21a8. Transcript *available at* http://www.cq.com/doc/congressionaltranscripts-3977668.

Senators Lieberman and Collins have criticized the training received by state and local law enforcement on similar grounds: "In addition, state and local law enforcement often have little or no guidance from the federal government on what counterterrorism training should entail. The result has been cases of trainers spewing inaccurate or even bigoted information to state and local law enforcement personnel, stigmatizing Muslim-Americans generally, and in effect, lending support to the false narrative that we are 'at war' with Islam. As we have stated in previous letters to this Administration, we have serious concerns that improper training may not be isolated occurrences and could be detrimental to our efforts to confront homegrown terrorism." Letter from Senators Joseph Lieberman and Susan Collins to Counterterrorism and Deputy National Security Advisor John Brennan, Sept. 12, 2011, *available at* http://www.hsgac.senate.gov/media/senators-urge-administration-to-address-internet-radicalization.

Finally, the Associated Press has uncovered documents showing that the New York City Police Department has been conducting widespread surveillance of mosques and Muslim-Americans: "Dozens of mosques and student groups have been infiltrated,

It is not clear whether the message being delivered to law enforcement by Washington about the prevalence and danger of an ideological domestic threat has contributed to such instances of law enforcement focusing on protected speech and religion rather than on terrorist activity. We have no in-depth analysis of the effects of the changes made since the terrible attacks of 9/11 intended to strengthen law enforcement's capabilities to prevent attacks or of the many promises that civil liberties and civil rights are being protected even while surveillance and other rules are watered down.

For example, amendments to the Attorney General's Guidelines governing FBI investigations now permit the FBI to gather vast amounts of data on Americans, including through "assessments" discussed in Lederman's paper. Those Guidelines were originally written to protect civil liberties by keeping the FBI focused on its mission: prevention and punishment of crimes. But FBI information collection is no longer confined to investigations of terrorist plots; it is now allowed for the ill-defined but extremely broad purpose of "foreign intelligence" gathering. Has the broadening of the crimes of material support and the use of material witness laws to detain suspects weakened the First Amendment protection for unpopular speech and religion?

Has the emphasis on a threat from "Islamist extremism" resulted in discriminatory law enforcement practices, including the profiling of people based on their religion or ethnicity in violation of the Constitution's promise of equal protection? And what about the concern about entrapment and the discriminatory use of sting operations targeted against individuals who are identified on the basis of their religious beliefs or political speech. That such operations have to date been upheld by courts operating under rulings that make entrapment a very difficult defense is no answer to the

---

and police have built detailed profiles of Moroccans, Egyptians, Albanians and other local ethnic groups. The NYPD surveillance extended outside New York City to neighboring New Jersey and Long Island and colleges across the Northeast." E. Sullivan, P. Yost, *Attorney General reviewing NYPD spying complaints*, ASSOCIATED PRESS, Feb. 29, 2012, *available at* http://ap.org/pages/about/whatsnew/wn_022911a.html.

concern that more Americans are worried that their government does not operate fairly and in accord with constitutional protections.

Have the lessened constraints on FBI investigations along with louder rhetoric about an insider threat from a particular religious ideology had the effect of creating misunderstandings and lack of consideration for the important Fourth and First Amendment values at stake in law enforcement investigations? Even if FBI leaders still understand the importance of constitutional limits, how are these changes understood by the thousands of officers and agents around the country, who are repeatedly told that their primary mission is to prevent another attack by "Islamist" radicals?

Negative Consequences of Singling Out and Labeling "Violent Islamist Extremism" Separate from Other Acts of Domestic Terrorism

As Lederman acknowledges in a footnote, the most deadly incident of homegrown terrorism in recent decades was Timothy McVeigh's murder of 165 Americans in Oklahoma City in 1995. Moreover, while one can argue in papers such as this that "Islamist" does not refer to the religion of Islam but to a perversion of the religion, in common parlance, the term itself does not necessarily imply any such limitation; to the contrary, it seems to imply a close relationship between the targeted extremism and Islam.

Michael Leiter, former Director of the National Counterterrorism Center, has cautioned about talking too much about the threat from al Qaeda and its adherents because of its tendency to glorify the group and make it seem "ten feet tall."[41] It also risks leaving Muslim and other communities feeling targeted, unfairly singled out, and persecuted. It risks alienating those communities that law enforcement says they most want to partner and collaborate with in preventing domestic terrorism.[42]

---

41. *See* M. Leiter, *The Changing Terrorist Threat and NCTC's Response*, Center for Strategic and International Studies, Dec. 1, 2010, *available at* http://csis.org/files/attachments/101202_leiter_transcript.pdf.

42. The administration's Strategic Implementation Plan acknowledges the "potential to do more harm than good" in the broader arena of countering violent

But one must also wonder about more subtle but equally harmful effects of this articulation. Does it encourage the latent tendency of politicians to scapegoat minorities and engage in ideological witch hunts? Does it contribute to the mistaken understanding by part of the public that the Constitution differentiates between some religions and others? It is having a less obvious but more insidious effect on the way law enforcement operates by encouraging focus on ideology rather than on actions and, in particular, on Islam? The use of bigoted training materials,[43] the idea that Islam is the enemy, and a misunderstanding of religious practices all may be encouraged by such language and focus. It also seems likely that such rhetoric may be responsible in part for the misunderstandings about constitutional limits on government surveillance or interference with religious practices and beliefs. Does it play a role in the decisions to send undercover informants into mosques or Muslim student groups with no predicate of any criminal activity? The strength of constitutional protections depends in part on local law enforcement's understanding and commitment to such protections. Is that understanding and commitment being undermined by leaders and politicians in Washington? While they insist that Islam is not the enemy, local experiences in the past few years suggest that local law enforcement may be receiving a different message.

Finally, it is important to examine whether the emphasis by government officials on the preeminent threat of violent Islamist extremism has resulted, even unintentionally, in encouraging the alarming rise in expressions of bias and hatred toward Islam and Muslims not only by fringe groups, but also by Washington politicians and pundits.[44]

---

extremism. EXECUTIVE OFFICE OF THE PRESIDENT, STRATEGIC IMPLEMENTATION PLAN FOR EMPOWERING LOCAL PARTNERS TO PREVENT VIOLENT EXTREMISM IN THE UNITED STATES, December 2011, at 3.

43. *See* n. 8, *supra.*

44. *See* Matt Duss, *Creeping Sharia 'Team B' Report Presented to Congress,* THINK PROGRESS, Sept. 15, 2010, at http://thinkprogress.org/security/2010/09/15/176274/creeping-sharia-team-b-report-presented-to-congress/.

Countering Violent Extremism: the Proper Role of the Government?

There is now a massive government effort underway to "counter violent extremism" including the administration's National Strategy issued this past year, congressional hearings, and, of course, law enforcement and intelligence efforts. As framed, this effort seems aimed at more ambitious ends than simply preventing violence; it also seeks to affect Americans' ideas and beliefs, whether or not they lead individuals to terrorist acts. In broadening the aim of government activity beyond the detection and prevention of terrorism, important questions are raised about the appropriate and effective uses of government, which, however, have received little attention.

In the United Kingdom, there was a massive government effort, the "Prevent" strategy, to counter Muslim extremism. Much of it involved identifying and funding "moderate Muslim" groups to act as a counterweight to the ideas of more "radical" groups.[45] Such an approach in the United States would, of course, run afoul of the First Amendment; the state has no business identifying religious groups as moderate or radical and then promoting one over the other. Doing so would be fundamentally at odds with our history and understanding of the proper relations between church, state, and individual.

The administration's National Strategy does not embrace the British model and it evidences appreciation for the difficult question of the proper role of government, beyond prevention of violence and the use of the bully pulpit by political leaders. At the same time, it is not clear whether the rules and structures adopted to implement this strategy will ameliorate rather than exacerbate the concerns outlined above.

Most striking perhaps is the failure to highlight the answer embedded in our constitutional structure: more speech. In a marketplace of ideas, advocacy of violence and intolerance will be defeated

---

45. For a critical appraisal of the Prevent program, *see* A. Kundnani, "Spooked! How not to prevent violent extremism," at http://www.srec.org.uk/Reports/59/Spooked!-How-not-to-prevent-violent-extremism-by-Arun-Kundnani.php.

by defense of constitutional, peaceful, and democratic ideals.[46] As former Counterterrorism Director Michael Leiter put it, demonstration of American resiliency will show the terrorists that they will not succeed.[47] Those who decry the ideological threat of al Qaeda or "radical Islam" show little confidence in that resilience or in our constitutional structures.

The Threat from Within: Reply to Kate Martin

Gordon Lederman

Kate Martin argues that Director of National Intelligence James Clapper, in his January 2012 annual threat assessment, did not include concerns that an American might self-segregate within his or her Internet community, radicalize to violent extremism, and then attempt to use modern technology to kill a large number of people. However, Director Clapper's threat assessment concerning chemical, biological, radiological, and nuclear threats (CBRN)—including passages quoted by Martin—is focused on the next year, and even then provides little reassurance.

- "[A] mass attack by foreign terrorist groups" involving CBRN is "unlikely in the next year." "Unlikely in the next year" would seem quantifiable as a 20 percent chance—hardly comfortable odds given the consequence of such an attack.
- U.S. Intelligence "worr[ies] about a limited CBR attack" within or outside of the United States in the next year because of interest by foreign terrorist groups.
- Later in his assessment, he notes that some homegrown violent extremists aspired to mass-casualty events but lacked the technical capability, except for two who were trained

---

46. Perhaps the best example of this was detailed in a story about a conservative Imam's conversations with young Muslims. *See* A. Elliot, *Why Yasir Qadhi Wants to Talk About Jihad*, N.Y. Times Magazine, March 17, 2011, *available at* http://www.nytimes.com/2011/03/20/magazine/mag-20Salafis-t.html?pagewanted=all.

47. *See* Leiter, n.8, *supra*.

outside of the United States and attempted mass-casualty explosive attacks within the United States.

- Finally, with respect to CBRN, U.S. Intelligence believes that lone homegrown violent extremists "are capable of conducting limited attacks in the next year" although the threat posed by anthrax is "low."[48]

Thus, the threat picture painted by Director Clapper *for only the next year* includes an approximately 20 percent chance of a major CBRN attack by a foreign terrorist group and a "worrisome" threat (however defined) of a limited such attack—possibly by a homegrown violent Islamist extremist who is trained abroad—and that homegrown violent extremists can perpetrate limited attacks aside from using anthrax.

Expanding beyond a one-year time horizon and taking a long-term view of the 21st-century insider threat requires looking at the vectors that affect the threat and points to the threat growing from "limited" to endangering millions of people. In his testimony, Director Clapper discusses various dynamics that could influence homegrown violent extremism, with two being that homegrown violent extremists will learn from previous plots, which seems highly plausible, and that they are using the Internet, which is already the case.[49] Moreover, in her essay, Martin does not confront the long-term diffusion of technology and expertise, particularly biological and cyber, which can be used to produce a weapon of mass destruction (WMD). We cannot assume that an individual with technical skills and access to WMD materials will be immune from radicalization within the United States or unavailable to train homegrown violent extremists abroad.

---

48. Statement by Director of National Intelligence James Clapper before the Senate Select Committee on Intelligence, Jan. 31, 2012, at 2, 4, *available at* http://www.dni.gov/testimonies/20120131_testimony_ata.pdf.

49. *Id.*, at 4. For an example of terrorist use of the Internet, *see* Senate Committee on Homeland Security and Governmental Affairs, Staff Report, *Zachary Chesser: A Case Study in Online Islamist Radicalization and Its Meaning for the Threat of Homegrown Terrorism* (2012), *available at* http://www.hsgac.senate.gov/imo/media/doc/CHESSER%20FINAL%20REPORT(1).pdf.

The cases that constitute the increased trend of homegrown violent Islamist extremist terrorism are few as compared to the millions
of patriotic Muslim-Americans, but that small number is only one
element of the calculation. The potential lethality—the consequence—
is another critical factor, as is intent—violent Islamist extremists'
desire to kill large numbers of Americans. The convergence of vectors over the next decade does, to use Martin's word, create a threat
of sufficient "magnitude" to affect government resource allocation
and organization. The issue is not whether a catastrophic insider attack is guaranteed, but rather that the probability of an attack is not
minimal, the consequences would be enormous, and various vectors
enhance the potential for an attack, thus justifying government's efforts to build a long-term defensive architecture, within Constitutional parameters, to prevent it.

Key components of this architecture include (1) building long-
term, trusted relationships with Muslim-American communities and
(2) enhancing law enforcement, intelligence, and other capabilities
such as by training government personnel. As the Obama Administration recognized in releasing its domestic community engagement
strategy to counter violent extremism, merely relying, as Martin suggests, on "a marketplace of ideas" is insufficient. But Martin correctly warns that some members of the public—or government personnel—might conflate Islam with violent Islamist extremism. Accordingly, government officials, as Presidents George W. Bush and
Barack Obama have done, must continuously articulate the distinction and highlight Muslim-Americans' contribution to America. In
addition, training of government personnel must meet the highest
standards of professionalism. News articles during the past year highlighted inaccurate and even bigoted training, conflating Islam and
violent Islamist extremism, by third-party trainers to state and local
officials. Governments need to be vigilant, not only because it is the
right course of action but also because negative publicity of such
training risks reinforcing the violent Islamist extremist narrative that
the West is at war with Islam and undermining the relationships that
governments should build with Muslim-American communities.

The Threat from Within: Reply to Gordon Lederman

Kate Martin

There is no doubt a continuing threat of terrorist attacks in the United States.[50] But the likelihood of a radicalized "Islamist" American using CBRN technology to kill millions of Americans is much less clear and will not be settled by this debate. The more immediate issue is whether looking at the potential of mass casualty terrorism through the lens of potential attacks by homegrown "Islamist extremists" rather than potential attacks by a broad range of actors risks missing important elements and wrongly emphasizing other elements. For example, wouldn't we be better served by an analysis that distinguished between terrorist attacks directed or inspired by al Qaeda and attacks carried out for other reasons, including, for example, potential attacks by Iranian agents. This approach would ensure incorporating an understanding of the larger context of al Qaeda's power and influence and, for example, an understanding of internal disagreements in al Qaeda on whether to focus future attacks on the continental United States. Such a more specific understanding and analysis of different groups and dynamics is necessary to fully evaluate potential risks, the "vectors" leading toward or away from such risks, and accordingly the best ways to meet those risks.[51] Shouldn't the possibility of a CBRN attack be examined from the perspective of a possible attack by many different actors, not just by a "violent Islamist" extremist? Wouldn't it be wiser to look more broadly at access to the necessary knowledge and materials to carry out such attacks—to include, for example, the possibility of a government scientist or a right-wing terrorist worried about Muslims taking over the country?

---

50. It is interesting, however, that, as one expert points out, the level of terrorist violence inside the U.S. since 9/11 has been appreciably less than it was in the 1960s and 1970s. Brian M. Jenkins, *Al Qaeda in Its Third Decade: Irreversible Decline or Imminent Victory?*, Occasional Paper, Rand Corp., 2012.

51. *Id.*

Apart from mass casualty attacks, most experts and even politicians in moments of honesty concede that it is simply not possible to stop all terrorist attacks. Thus, we need government efforts to prepare for such attacks as well as to prevent them. This is especially important as the terrorists' aim to create fear and overreaction can be defeated by a smart and effective response.

While there is general agreement about the existence of a problem, serious questions remain whether current policy discussions in Washington adequately consider the real benefits versus costs of certain government counterterrorism resources and architecture. For example, it has become clear that there is an inherent tension between the ways in which law enforcement/intelligence capabilities are being expanded and the stated goal of building long-term relationships of trust with Muslim-American communities; law enforcement's current claim that it may collect vast amounts of information about such communities—without any individualized suspicion of wrongdoing—contributes to that tension. The current controversy concerning blanket NYPD surveillance of Muslim communities is evidence that this tension has not been adequately addressed, much less resolved.

In addition, while Presidents Obama and Bush have publicly repeated that the United States is not at war with Islam, there has been very little examination of whether that understanding has been internalized and operationalized by the thousands of domestic law enforcement agencies now tasked with counterterrorism. The recent revelations of widespread bigoted training materials are disturbing not only as Lederman points out because they are wrong and likely to undermine community trust, but also because they raise serious questions about the possible injection of stereotypes, prejudice, and unwarranted fears among law enforcement and intelligence personnel.

Nor is this problem confined to government personnel. Sadly, while there have been vehement condemnations of the rise in public expressions of bigotry, there has been little examination of the possible, though unintended, consequences of government pronouncements about the threat of "Islamist" terrorism on public attitudes toward Muslims. Indeed, one of the objections to the term *violent Islamist extremism* is that it obscures the difference between the religion of Islam and terrorist ideologies, playing into public fears and bigotry.

Finally, while reliance on the "marketplace of ideas" as an important strategy to counter extremist violence is usually dismissed as simplistic or naïve although not by Lederman, there is certainly historical evidence of the strength of ideals of democracy and individual liberty to defeat whatever appeal terrorists offer, most recently, of course, in the events of the Arab Spring. When worrying about the appeal of terrorist violence to Americans, it is crucial not to overlook the importance of individuals feeling free to contemplate, discuss, and publicly advocate for whatever ideas they wish as the alternative to engaging in terrorist crimes.

Chapter Three
Interrogations

Norman Abrams
(Proposing a "Cabined Exception" for terrorism interrogations)

Christopher Slobogin
(Opposing a "Cabined Exception" for terrorism interrogations)

Introduction

... [T]the threat posed by terrorist organizations and the
nature of their attacks—which can include multiple accom-
plices and interconnected plots—creates fundamentally different public safety concerns than traditional criminal cases.[1]

—Matthew Miller
Spokesman for the U.S. Justice Department
May 24, 2011

Treating terrorism suspects as if they're archvillains for whom
rights are too dangerous a thing to contemplate plays directly into their hands. What is terrorism if it's not the efforts of a small group of people to hold a much larger group
hostage to fear? When we act as if these terrorism suspects
are such a threat that the normal rules cannot be applied to

---

1. Evan Perez, *Rights Are Curtailed for Terror Suspects*, WALL ST. J., March
24, 2011, *available at* http://online.wsj.com/article/SB100014240527487040
5020457621897065211 9898.html#articleTabs%3Darticle.

them without putting our entire system of justice in jeopardy, we've essentially done their work for them.[2]

—Michael German
Policy Counsel for National Security, ACLU
May 21, 2010

The Case for a Cabined Exception to Coerced Confession Doctrine in Civilian Terrorism Prosecutions

Norman Abrams

Introduction

his paper presents the case for recognition of an exception—a cabined exception—to coerced confession rules governing the admissibility of interrogation statements in civilian terrorism prosecutions. It has been many years since the Supreme Court has addressed coerced confession issues in a significant way.[3] Accordingly, it might be thought: Why address an issue that appears to lie in a judicial backwater?

A number of factors account for the absence, for a long time, of high court interest in this subject. In criminal prosecutions in civilian federal courts regarding interrogation statements, *Miranda*[4] has long occupied center stage. While issues of voluntariness have occasionally been addressed in recent years by the Supreme Court and by the lower federal courts, most such instances have involved questions of whether waiver of the *Miranda* warnings requirement was voluntary.[5] And where an interrogation has followed voluntary waiver of

---

2. Michael German. "Keep Constitution Intact When Interrogating Terrorists," *ACLU* (http://www.aclu.org/blog/national-security/keep-constitution-intact-when-interrogating-terrorists).

3. *See* Catherine Hancock, *Due Process Before* Miranda, 70 TUL. L. REV. 2195 (1996). *Also see*, however, Colorado v. Connelly 479 U.S. 157 (1986).

4. Miranda v. Arizona, 384 U.S. 436 (1966).

5. *See* North Carolina v. Butler, 441 U.S. 369 (1979). *See generally* Welsh S. White, Miranda*'s Failure to Restrain Pernicious Interrogation Practices*, 99 MICH. L. REV. 1211 (2001).

Miranda rights, defendants have had a difficult time proving that the statements they made were coerced.[6]

It is noteworthy, too, that in the post–9/11 period, military commission prosecutions have not as yet led to many judicial decisions on coerced confessions. Of course, the main explanation for that fact (despite allegations of torture) is that not very many military commission prosecutions have been tried to completion.[7] More recently, the Obama administration attempted to switch the focus in the military detainee context away from torture by prohibiting in any armed conflict the use in interrogation of any method that "is not authorized by and listed in the Army Field Manual. . . ."[8]

The absence of significant court decisions relating to coerced confession issues in terrorism prosecutions on both the civilian and military side is likely to change. On the military side, as more cases are being tried, coerced confession issues respecting past interrogations are likely to arise frequently. For the future, because the *Army Field Manual (AFM)* rules are to be applied by interrogators, questions likely to arise in subsequent trials will be: (a) whether some of the methods permitted under the *Manual* are consistent with the confessions provisions of the Military Commissions Act of 2009,[9] (b) whether existing Supreme Court coerced confession doctrine[10] is applicable in military

---

6. *See* Welsh S. White, op. cit., *supra* note 5 at 1220.

7. On the other hand, the admissibility of statements allegedly coerced has been the subject of frequent and intensive litigation in the habeas corpus actions brought by Guantanamo detainees in the wake of the decision in *Boumediene v. Bush*, 553 U.S. 723 (2008). *See* Benjamin Wittes, Robert M. Chesney & Larkin Reynolds, The Emerging Law of Detention 2.0 (Harvard Law School, Brookings, May 2011) for a detailed description of the judicial treatment of the allegations of coercion by the detainees in the habeas litigation. Potentially, the treatment of coerced confession issues in the habeas cases may influence future decisions in military commission trials.

8. Exec. Order, Ensuring Lawful Interrogations, Jan. 22, 2009. *Also see* U.S. Dep't of the Army, Field Manual on Interrogation.

9. Sec. 948r, Military Commissions Act of 2009.

10. Thus, the AFM rules permit some forms of deception. A number of Supreme Court decisions have held a confession to have been involuntary in instances where an important element underpinning the Court's conclusion was the fact that the confession was obtained by some kind of deception. *See, e.g.*, Spano v. New York, 360 U.S. 315 (1959); Leyra v. Denno, 347 U.S. 556 (1954).

commission proceedings,[11] and (c) if so, whether application of *AFM*-approved interrogation methods is consistent with that doctrine.[12] Adoption of the *AFM* rules may be effective to prohibit torture by the military or the CIA,[13] but it may turn out not to have been an especially good way to ensure that interrogations are lawful under the applicable constitutional coerced confession doctrine.

More important to the subject of this paper, in connection with civilian enforcement against terrorism, new FBI interrogation guidelines[14] have been adopted, which, subject to being upheld by the courts, permit exigent circumstance terrorism interrogations without *Miranda* warnings. These guidelines should eliminate in many terrorism interrogations the issue of the voluntariness of a *Miranda* waiver and open the door instead to a direct focus on whether the statements were lawfully obtained under coerced confession doctrine.

Accordingly, the application of coerced confession doctrine in civilian terrorism investigations and trials (and in military cases, too) in the future is likely to be an issue that is addressed more frequently in the federal courts and, in due time, may reach the Supreme Court.

The Case for Recognition of a Cabined Exception

There are a number of grounds, which, taken together, support the case for recognizing a cabined exception to coerced confession doctrine.

---

11. In this connection, consider the implications of *Boumediene v. Bush*, 553 U.S. 723 (2008).

12. *See, e.g.*, note 10 *supra*.

13. The Executive Order requiring that the AFM rules be followed does not seem to have been intended to include the FBI among the agencies to which it applies, though the relevant language is hardly a model of clarity.

14. The existence of these Guidelines (which had been promulgated on Oct. 21, 2010) was first publicly disclosed in *The New York Times*, which published a copy on March 25, 2011.

1. Exceptions to Other Rules of Constitutional Evidentiary Admissibility

The fact that other closely related exceptions to constitutional rules of evidentiary admissibility have already been recognized or proposed lends credence to the idea of formulating an exception in the confessions arena. Indeed, it can be argued that recognition of exceptions to existing constitutional evidentiary admissibility rules in civilian terrorism prosecutions is a developing phenomenon.[15] Special rules have long been applied, for example, in connection with judicial approval of electronic eavesdropping and searches in terrorism[16] contexts, under the heading of foreign intelligence;[17] in recent years, a related special rule has been used by a couple of federal courts in compulsory process contexts.[18] Also, arguments have been made by scholars[19] for extending to terrorism cases the aforementioned "dispensing with *Miranda* warnings in exigent circumstances"[20] exception (an approach that has been implemented through adoption of the aforementioned FBI Guidelines). Similarly, it has been argued by this author that a similar extension can be adopted for the exigent-circumstance exception to confrontation

---

15. *See generally* Norman Abrams, *Terrorism Prosecutions in Federal Court: Exceptions to Constitutional Evidence Rules and the Development of a Cabined Exception for Coerced Confessions*, UCLA School of Law Research Paper No. 11-14 (2011) (http://papers.ssrn.com/sol3/papers.cfm?abstract_id=1846963).

16. *See, e.g.*, the definition of "agent of a foreign power" in the FISA statute in § 1801(b)(1)(C).

17. Foreign Intelligence Surveillance Act of 1978, 50 U.S.C. § 801 *et seq.* The Supreme Court in *United States v. United States District Court*, 407 U.S. 297 (1972), a domestic terrorism case, indicated that Congress could legislate different standards for search warrants to obtain intelligence in domestic security cases.

18. *See* United States v. Moussaoui, 382 F.3d 453 (4th Cir. 2004); *see also* United States v. Paracha, 2006 WL 12768 (S.D.N.Y. 2006), *aff'd*, 2008 U.S. App. LEXIS 12937.

19. Paul Cassell, *Time to Codify a* Miranda *Exception for Terrorists?*, VOLOKH CONSPIRACY (Oct. 21, 2010), at http://volokh.com/2010/ 10/21/time-to-codify-a-miranda-exception-for-terrorists/.

20. But not without other scholars expressing disagreement. *See, e.g.*, Amos N. Guiora, *Relearning Lessons of History:* Miranda *and Counterterrorism*, 71 LA. L. REV. 1147 (2011).

recently recognized by the Supreme Court.[21] Given such exceptions in terrorism cases, already recognized or proposed, it seems natural to ask whether a similar exception might be applied to the existing coerced confession doctrine.

2. The *New York v. Quarles* Social Cost Argument

The strongest model for the development of such an exception is the decision in *New York v. Quarles*[22] that ruled in a particular exigent circumstance that *Miranda* warnings were not required before questions were put to the suspect. The Court applied a social cost calculus in reaching its conclusion, namely, there might have been an unacceptable social cost if the *Miranda* warnings had been given and, as a result, information about a missing gun had not been forthcoming from the suspect.[23]

3. The Social Cost Issue in an Investigation-of-Terrorism Context

Application of a social cost argument in an exigent circumstance–terrorism context is likely to be based on a much greater unacceptable cost than that which was involved in *Quarles*. In that case, the cost of not getting the needed information from the suspect—where the gun was—would be the risk to human life if a single gun fell into the wrong hands. In a terrorism investigation context, the social cost if the needed information is not obtained could be the occurrence of a catastrophic terrorist act that might cost dozens, hundreds, or even thousands of lives. A strong argument can be made in favor of adopting a rule permitting the use of some limited coercive techniques, beyond methods permitted in ordinary crime cases, if the gain may be to obtain information from the suspect about an impending terrorism event. Of course, the coercive techniques used against the suspect will be a greater intrusion than that involved under the *Miranda* exception, but

21. *See* Norman Abrams, op. cit., *supra* note 15.
22. 467 U.S. 649 (1984).
23. The Department of Justice relied primarily on *Quarles* in promulgating the FBI interrogation guidelines, referred to *supra*, at note 14.

the somewhat greater intrusion arguably is counterbalanced by the potentially much greater social cost thereby avoided.

4. Can the Social Cost Calculation Be Used to Justify Torture?

A problem with applying a social cost argument in support of recognizing an exception to the coerced confession doctrine is the concern that it might be used to justify extreme coercive interrogation techniques to the point where even torture might be utilized. The use of torture or other extreme methods of interrogation should be unlawful under all circumstances. But there are interrogation techniques, not remotely comparable to torture and nevertheless unlawful under existing Supreme Court coerced confession case law,[24] which government agents should be allowed to use in trying to obtain intelligence about an impending terrorism event. *The way to achieve this outcome is to recognize a cabined exception to existing coerced conf ession doctrine*, which would only go so far as to permit the use of non-extreme interrogation methods, including some that would be unlawful under Supreme Court precedents derived from ordinary crime cases.

5. Is There Empirical Evidence That Supports a Need for This Exception?

Is there a demonstrated need to increase the range of methods that the FBI can use in interrogating suspects? Is there any empirical evidence to support such a claim? Justification for adopting a limited exception to existing coerced confession doctrine is not based on any claim of demonstrated need. Rather, the argument is that the prevailing doctrine is more restrictive than necessary where the purpose of the interrogation is to obtain intelligence about an impending terrorism event. Avoiding the harm of a catastrophe is worth additional intrusions on the individual that do not involve the use of extreme interrogation

---

24. *See, e.g.*, the type of deception used by police interrogators in *Spano v. New York*, 360 U.S. 315 (1959) and *Leyra v. Denno*, 347 U.S. 556 (1954).

methods. Even a small modification of confession rules will serve to give the FBI additional flexibility in conducting an interrogation.

6. Providing Guidance to FBI Field Agents

Under the current state of the law, FBI field agents are likely to be uncertain what confession rules will be upheld by the courts. They need clear guidance on this topic. It is desirable to establish clearly drawn rules and, in the course of doing so, to give interrogators somewhat more leeway than they have under prevailing Supreme Court precedents. For example, the *AFM* rules or a similar body of rules to be formulated may be used as a set of guidelines for both the courts and field agents.

7. Avoiding a Further Muddling of the Doctrine

Absent such an exception, the doctrine in the terrorism arena is likely to become even more muddled than in ordinary crime cases and that kind of development could begin to infect ordinary criminal cases. If such an exception is not adopted, many courts are likely to apply a more flexible approach in addressing coerced confession issues where the interrogation is designed to elicit information about an impending terrorist act,[25] but not all courts will be willing to overlook the strictness of ordinary crime coerced confession rules. As a result, absent a recognized, clearly drawn exception, more than a normal modicum of inconsistency is likely to arise in the case law, which will put stress on coerced confession doctrine in terrorism cases and in ordinary crime investigations, too.

8. The Somewhat Reduced Risk of Unreliability in Statements Obtained through Terrorism Intelligence Interrogations

Still another reason why such an exception is warranted arises out of one of the concerns underlying the traditional coerced confession

---

25. If the defendant appears to be a dangerous terrorist, a judge is likely to be very reluctant to release him or her back into the community.

rules—the fact that extreme coercive interrogation methods often lead to unreliable confessions: The suspect will say anything to get the pain to stop. Even a non-extreme interrogation may produce untrustworthy statements where the interrogation is conducted by law enforcement agents trying to obtain information to be used in prosecuting the suspect. In such a context, the questioning agent is strongly motivated to direct questions in a way that maximizes the possibility of obtaining a confession that can be used to convict the suspect. This is a kind of bias that can substantially influence the form and content of the interrogation and thereby increase the risk that the answers given are unreliable.

What is contemplated under the proposed exception is an interrogation directed to obtaining terrorism intelligence: Questions are directed not to obtaining a confession usable in prosecuting the suspect but rather to obtaining accurate information that can help to prevent terrorist acts. In such a context, there is a premium on accuracy, not conviction. Questioning terrorists for intelligence purposes thus modulates one of the factors[26] that ordinarily contribute to a risk of unreliability in law enforcement interrogations.

9. Fifth and Fourteenth Amendment Underpinnings of the Coerced Confession Doctrine

How will such a cabined exception fit with the existing coerced confession doctrine? Initially, the idea of applying an exception via the Fifth Amendment underpinning of the traditional doctrine[27] seems sharply inconsistent with an idea expressed in opinions of some Supreme Court justices—that the Fifth Amendment does not brook any exceptions.[28] *New York v. Quarles* carved out an exception to *Miranda* which was arguably based on the Fifth, but the majority opinion in

---

26. It surely does not eliminate, but arguably reduces to an indeterminate extent, the risk of unreliability.

27. *See* Bram v. United States, 168 U.S. 532 (1897).

28. *See* Justice Scalia dissenting in *Dickerson v. United States*, 530 U.S. 428, 453 (2000), *and see* Justice Marshall dissenting in *New York v. Quarles*, 467 U.S. 649, 688 (1984).

Quarles characterized *Miranda* as a prophylactic rule not of constitutional dimension. Subsequently, however, *Dickerson v. United States*[29] reinstated the view that *Miranda* is a constitutional rule while expressly preserving the *Quarles* exception to *Miranda*. So much for the notion that the Fifth Amendment is intolerant of exceptions.

The view that coerced confession doctrine is based in the Due Process Clauses of the Fourteenth and Fifth Amendments provides a more receptive path to take in support of the exception. To begin with, the notion of due process seems flexible enough to include taking into account the reasons for using the interrogation techniques in question as well as what those techniques are. Thus, weighing in the balance the justification for the intelligence interrogation and the relative "coerciveness" of the interrogation techniques used against the potential social cost of not permitting certain techniques to be used seems well within a reasonable application of a due process calculus. To be sure, there is very limited Supreme Court authority[30] to cite in support of such a conclusion, and nothing directly in point. It is fair to say, however, that nothing in existing doctrine under either the Fourteenth or Fifth Amendments stands as a clear obstacle to the development of such an exception.

Conclusion

In crafting and implementing an exception, numerous issues will need to be addressed: determining the terms in which the cabined exception is to be formulated; where and how to draw the line that limits the exception to non-extreme interrogation techniques; how to

---

29. 530 U.S. 428 (2000).

30. *See, e.g.*, Chavez v. Martinez, 538 U.S. 760 (2003), where the lawfulness of an interrogation was at issue in a civil rights lawsuit. Justice Thomas, joined by Chief Justice Rehnquist and Justice Scalia in discussing whether due process has been violated, inter alia, stated: "Moreover, the need to investigate whether there had been police misconduct constituted a justifiable government interest given the risk that key evidence would have been lost if Martinez had died without the authorities ever hearing his side of the story. . . ."

define exigency;[31] how to prevent expansion of the exception through interpretative erosion of its terms or extension to other crime areas; and what should be the first step(s) on the path to the recognition of such an exception.[32] Great care will need to be taken in both the formulation and implementation of the exception.

In view of an upcoming likely increase in the incidence of confessions litigation in terrorism cases, developing a cabined, exigent circumstance exception to coerced confession rules now would be both a timely and desirable addition to the existing and proposed array of exceptions to constitutional evidence rules applicable in terrorism prosecutions.

An Unneeded and Dangerous Exception

Christopher Slobogin

Introduction

In the wake of 9/11, the temptation to use high-pressure interrogation techniques on individuals suspected of being terrorists skyrocketed. Professor Abrams argues that our government should be permitted to give in to that temptation, at least up to a point. He contends that: (1) there are interrogation practices that fall short of torture and yet are more coercive than practices currently permitted under American law; (2) these intermediate techniques are important tools for combating terrorism; and (3) courts can (as a matter of precedent) and should (as a matter of policy) develop a "cabined exception" to the traditional prohibition on coerced confessions so

---

31. The FBI guidelines address some of the issues that will need to be treated in implementing the guidelines. In a lengthier paper on the subject, I have dealt with many of these issues in more detail. *See* Norman Abrams, op. cit., *supra* note 15.

32. For example: Enacting legislation on the subject? Promulgation of internal governmental guidelines like those adopted by the FBI to implement a *Quarles*-like exception to *Miranda*? Interrogations by the FBI designed to test the issue without first formulating guidelines? Waiting for the right case to pose the issue directly to the courts?

that government officials may legally use these techniques. In this response, I comment on each of these propositions.

The Scope of a Cabined Exception to the Coerced Confession Rule

Outside of one ambiguous reference noted below, Professor Abrams does not provide any specific examples of procedures that would fall under his cabined exception. Instead, he suggests that the rules found in the *Army Field Manual* and the FBI's new interrogation Guidelines would be a good baseline for developing techniques that, while impermissible under the Supreme Court's current interpretation of the Fifth Amendment and the Due Process Clause, would be permissible in national security investigations. To get a sense of how his cabined exception would work, some understanding of those rules is necessary.

The *Army Field Manual*'s interrogation provisions, which overlap considerably with the FBI Guidelines (to the extent the latter provides specifics), prohibit "torture or highly degrading acts," a phrase meant to include forcing the detainee to stand naked or perform sexual acts, hooding or covering the eyes of the detainee, beating or shocking the detainee, or depriving the detainee of necessary food and water.[33] Presumably, given his reference to the *Manual* as a guidepost, Professor Abrams would not permit these techniques, either. However, as of 2006, the *Manual* no longer explicitly prohibits two techniques that it at one time banned—putting the suspect in stress positions and sleep deprivation.[34] The *Manual* also does not explicitly prohibit questioning of individuals who are in extreme physical pain because of an earlier injury, nor does it bar suggestions to examinees that a failure to cooperate will increase the chance of subsequent injury by third parties. So, would Professor Abrams' cabined exception permit use of the latter techniques even if, as it

---

33. U.S. ARMY FIELD MANUAL ON INTERROGATION 5–21.
34. *Id.* at M-9.

seems quite likely, current Supreme Court doctrine categorically rejects such practices?[35]

The answer is not clear. But Professor Abrams appears to shy away from defending coercion induced by the threat or fact of physical harm directed at the suspect. Rather, the only specific examples he gives (in a note[36]) of techniques that he might permit despite their rejection by the Supreme Court come from *Spano v. New York*[37] and *Leyra v. Denno*.[38] In both cases, the police relied on psychological rather than physical coercion, principally by lying to the suspect. In *Spano,* interrogators made use of a police officer who was a friend of Spano's and whom Spano had called in a panic shortly after the alleged crime; under the interrogators' direction, this officer told Spano, falsely, that the officer was "in trouble" with his superiors because of Spano's phone call and that if he lost his job his pregnant wife and kids would suffer as well. In *Leyra,* a psychiatrist posed as a medical doctor brought in to treat Leyra's sinus condition. He then repeatedly told Leyra "how much he wanted to and could help him, how bad it would be for petitioner if he did not confess, and how much better he would feel, and how much lighter and easier it would be on him if he would just unbosom himself to the doctor."[39] In both cases, the Supreme Court found the confessions to be involuntary under the Due Process Clause.

Professor Abrams suggests, without declaring outright, that the deceptive practices in *Spano* and *Leyra* are illustrative of interrogation techniques he would permit in national security situations under his cabined exception. But it is not clear that an exception is needed to

---

35. *See* Ashcraft v. Tennessee, 322 U.S. 143 (1944) (confession coerced when police questioned suspect continuously for 36 hours without allowing him rest or sleep); Mincey v. Arizona, 437 U.S. 385 (1978) (confession coerced by questioning of a suspect while he was in the hospital receiving treatment for a gunshot wound and in "unbearable" pain); Arizona v. Fulminante, 499 U.S. 279 (1991) (confession coerced when government informant promised to protect suspect from "tough treatment" by other prison inmates only if suspect talked). *But see infra* note 44.

36. *See* Abrams, note 24.

37. 360 U.S. 315 (1959).

38. 347 U.S. 556 (1954).

39. *Id.* at 559–60.

accommodate that goal. Both *Spano* and *Leyra* were decided in the 1950s. Since that time, the Supreme Court has explicitly or implicitly sanctioned police failure to correct misunderstandings about the right to remain silent and the right to counsel,[40] permitted the police to suggest, as in *Spano,* that a confession will prevent harm to a third party,[41] and blinked at police lies about fingerprint evidence and a codefendant's confession.[42] Most lower courts have found nothing un-constitutional about other types of lies,[43] and in particular have been unmoved by the "pretended friend" technique, in which, as in *Leyra,* government agents pose as caring individuals whose only goal is to help the suspect.[44] Several lower courts have even sanctioned police

---

40. North Carolina v. Butler, 441 U.S. 369 (1979) (finding admissible a confession by a suspect who believed that only written confessions would be admissible); Colorado v. Spring, 479 U.S. 564 (1987) (holding admissible a confession by a suspect who believed that once he started talking he could not cut off questioning); Connecticut v. Barrett, 4779 U.S. 523 (1987) (holding admissible a confession by a suspect who believed he did not need an attorney as long as his statements were not reduced to writing).

41. Rhode Island v. Innis, 446 U.S. 291 (1980) (holding admissible a confession by a suspect who was told that if he did not reveal the location of the gun used in the crime, a child might find it and hurt herself).

42. Mathiason v. Oregon, 429 492 (1977) (strongly suggesting that lying about finding the suspect's fingerprints at the scene of the crime was not "relevant" to the admissibility issue); Frazier v. Cupp, 360 U.S. 315 (1959) (holding admissible a confession by a suspect who was told, falsely, that his co-defendant had just confessed).

43. *See* Paul Marcus, *It's Not Just About Miranda: Determining the Voluntariness of Confessions in Criminal Prosecutions,* 40 VAL. U. L. REV. 601, 612–13 (2006) (noting that courts have permitted lies about "witnesses against the defendant, earlier statements by a now-deceased victim, an accomplice's willingness to testify, whether the victim had survived an assault, 'scientific' evidence available, including DNA and fingerprint evidence, and the degree to which the investigating officer identified and sympathized with the defendant.").

44. *Id.* at 623 ("There are many cases in which confessions are found to be voluntary based upon a variety of promises made, including vague guarantees that the defendant will receive better treatment if she confesses, offers of more lenient punishment for the suspect, assurances of lesser charges being prosecuted if the individual confesses, and the receipt of medical treatment if she makes an incriminating statement."). One of the most dramatic examples comes from *Miller v. Fenton,* 796 F.2d 598 (3d Cir. 1986), where the detective repeatedly claimed he was the "brother" of the suspect and that psychiatric help would be forthcoming upon a confession.

threats.[45] Not all of these latter decisions are necessarily consistent with Supreme Court case law. But, as I have argued elsewhere, if the police avoid direct threats, promises that an officer is not empowered to make, or deception about the scope of one's right to remain silent or right to counsel, police ruses are not coercive and should be considered constitutionally permissible under current doctrine.[46]

The upshot of all this is that if Professor Abrams is principally interested in allowing interrogators to deceive their sources, he does not need an exception to current law. The Due Process Clause already grants interrogators considerable leeway in interrogating suspects. Only if Professor Abrams wants interrogators to be able to resort to physical coercion, threats of physical harm, or unauthorized promises would a cabined exception be necessary as a legal matter.

The Practical Necessity for a Cabined Exception

Let us assume, however, that deception of the type involved in *Spano* and *Leyra* is still unconstitutional or that it is not but that Professor Abrams is willing to permit at least some types of physical coercion or threats or promises that clearly are unconstitutional at the present time. Although that would mean a cabined exception to current doctrine is necessary as a legal matter, it does not mean it is necessary as a practical matter. As many have pointed out, extraordinary interrogation techniques are not a particularly effective method of obtaining information from suspected terrorists.

One of the strongest advocates for the latter point of view is former FBI interrogator Ali Soufan.[47] He points out that Abu Jandal, Osama

---

45. Marcus, supra note 43, at 619–20 ("Confessions have been found to be voluntary even with stark threats such as the following: a refusal to offer protection though a credible danger of violence existed, a statement that the police will arrest the defendant's wife, a threat to incarcerate the suspect for an extended period of time, a threat to possibly cause harm to the defendant, or a comment about removing a child from the defendant's family.").

46. Christopher Slobogin, *Lying and Confessing*, 39 TEXAS TECH. L. REV. 1275 (2007).

47. Bobby Ghosh, *After Waterboarding: How to Make Terrorists Talk?*, TIME MAGAZINE, June 8, 2009, *available at* http://www.time.com/time/magazine/article/0,9171,1901491,00.html.

Bin Laden's bodyguard and a figure so intimidating that guards wore masks when they interacted with him, ended up providing highly useful information after weeks of resistance and hostility, not because of tough techniques but because his interrogators finally treated him with enough respect to get him talking and then used "sleight of hand" to seal the deal. Soufan also contends that al Qaeda operative Abu Zubaydah gave up the identities of Khalid Sheikh Mohammed (the mastermind of the 9/11 attacks) and of "dirty bomber" Jose Padilla through a combination of "guile and graft." Several other military interrogators agree with Soufan that "the best way to get intelligence from even the most recalcitrant subject is to apply the subtle arts of interrogation rather than the blunt instruments of torture."[48]

Some who have researched interrogations believe that even "guile" is not necessary to get information out of most domestic suspects.[49] But even if terrorists are harder nuts to crack than the typical street criminal, they may disclose useful information prompted by nothing more than expert application of "minimization techniques," which involve empathizing with the subject and playing down the seriousness of his or her situation. Research indicates that this tactic, which is clearly constitutional, is extremely effective at getting people to talk.[50]

There is also the possibility that more strenuous techniques will significantly increase the unreliability of any information obtained. Professor Abrams minimizes this possibility on the ground that, in contrast to domestic police whose eagerness to obtain a conviction can lead them to overreach, national security interrogators are not

---

48. *Id.*

49. For instance, in both Great Britain and Australia police have moved toward "investigative interviewing," which begins with open-ended questions and then proceeds to confrontation with non-manufactured evidence if the suspect's statements are inconsistent with it. At least one study claims that confession rates remain high using this technique. DAVID DIXON, INTERROGATING MYTHS: A COMPARATIVE STUDY OF PRACTICES, RESEARCH, AND REGULATION, *available at* http://ssrn.com/abstract=168935812.

50. *See* Melissa Roussano et al., *Investigating True and False Confessions with a Novel Experimental Paradigm*, 16 PSYCHOL. SCI. 481 (2006) (finding that minimization techniques increased confessions from guilty subjects from 46% to 81%, although they also produced confessions from 18% of the "innocent" suspects).

interested in prosecution but only in obtaining information that can save lives.[51] In many cases involving alleged terrorists, however, prosecution in a criminal or military court is very much a possibility.[52] Furthermore, the pressure to obtain "good" information about a terrorist attack might create the same bad incentives that a desire for usable evidence in a civilian criminal prosecution does. Cases like those of Haji Pacha Wazir, Osama Bin Laden's personal banker who was subjected to detention and interrogation for eight years before being released, suggest that national security investigators are just as subject to tunnel vision and abuse of power as are domestic police.[53]

In short, techniques beyond those already permitted by the Constitution may not be necessary to protect the country in national security situations, and may even be counterproductive. At the least, use of more serious pressure tactics should be prohibited unless there is strong proof both that the individual has important information and that less strenuous tactics have failed. Even that limited approach is problematic, however, for reasons that are best explored in connection with debunking Professor Abrams' third proposition, which is that a cabined exception is a good idea under the conditions he posits.

The Provenance and Dangers of a Cabined Exception

To this point the argument has been that a cabined exception to the coerced exception requirement is not necessary either as a legal matter (given the constitutionality of interrogation deception) or as a practical matter (given the ease of obtaining information through non-coercive means). But even if those arguments are wrong, a cabined exception is not justifiable on legal grounds, nor is it a good idea on policy grounds. Contrary to Professor Abrams' claim, current precedent does not provide even a weak basis for a cabined

---

51. Abrams at 48, Argument 7.

52. *See* Mark Mazzetti, *Pentagon Revises Its Rules on Prosecution of Terrorists*, N.Y. TIMES, Jan. 7, 2007 (indicating that the military believes that 60 to 80 of the 395 individuals in Guantanamo could be tried on war crimes charges).

53. *See* Jason Leopold, *CIA Kidnapped, Tortured "the Wrong Guy," Says Former Agency Operative Glenn Carle*, TRUTHOUT, Oct. 23, 2011, *available at*

exception to the coerced confession rule. In any event, recognition of such an exception would be a troubling development even if it did occasionally provide useful information.

The Supreme Court decision that comes closest to providing license for a cabined exception to the coerced confession prohibition is, as Professor Abrams indicates, *New York v. Quarles*.[54] In that case, the Court recognized a "public safety" exception to the general rule that police must give a suspect subjected to custodial interrogation the famous *Miranda* warnings. At first glance, the *Quarles* holding would seem to be a perfect justification for relaxing restrictions in the national security context, where public safety is surely at risk.[55] But several reasons counsel against such a move. First, *Quarles* is only an exception to *Miranda*'s prophylactic warning requirement, not to the ban on coerced confessions. Indeed, in *Quarles* itself, the Court made clear that a claim of undue coercion would lead to exclusion even in a public safety situation.[56] Second, the *Quarles* rule only applies when public safety is imminently threatened; in *Quarles,* for instance, the police were looking for a gun that had just been discarded. Most interrogations of suspected terrorists are conducted in non-exigent circumstances, for the purpose of gathering intelligence or eventually prosecuting the source. Only if a situation arose in which an alleged terrorist were thought to have information about an immediate threat—the proverbial ticking bomb scenario—might danger to the public of the type contemplated in *Quarles* be present.

Even in the latter situation, however, the best approach as a matter of policy is to adhere to a ban on coerced interrogation, which should include a prohibition on the "softer" yet unconstitutional type of coercion apparently contemplated by Professor Abrams. As Judge Richard A. Posner has said about torture, "[h]aving been regularized, the practice will become regular. Better to leave in place the

---

54. 467 U.S. 649 (1984).

55. For an argument to this effect that is even more explicit than Professor Abrams', *see* Joanna Wright, *Mirandizing Terrorists? An Empirical Analysis of the Public Safety Exception*, 111 COLUM. L. REV. 1296 (2011).

56. 467 U.S. 655 n.5.

formal and customary prohibitions, but with the understanding that they will not be enforced in extreme circumstances."[57]

If an exception to the ban on coercion were recognized, however cabined it may be in theory, pressure tactics will become routine practice not only in cases involving alleged terrorists, but also will eventually find their way into interrogations of anyone *connected* with a terrorist organization, then anyone associated with "narco-terrorism,"[58] and finally anyone suspected of any crime. We've seen similar expansionary tendencies in other areas of the law. For instance, the Foreign Intelligence Surveillance Act, initially limited to wiretaps where national security was the "primary" purpose, now permits surveillance when national security is a "significant" government goal.[59] Even more insidious mission creep has occurred in connection with the fusion centers that have replaced the Department of Defense's Total Information Program; once confined to data-mining records for evidence of terrorists threats, they now routinely collect information about illegal immigrants and deadbeat dads.[60] It is much better to have rules that apply across the board to all interrogations, with the incentive to cabin them that universal application brings, than to try to limit the rules to a single kind of crime. Give government officials an inch . . . .

The Cabined Exception Proposal: A Reply to Professor Slobogin

Norman Abrams

Stripped down to essentials, Professor Slobogin's essay advances two sets of arguments against the cabined exception proposal that merit

---

57. Richard A. Posner, *The Best Offense*, New Republic, Sept. 2, 2002, at 28.

58. Amy Zalman, Narcoterrorism Evolves in the War on Terror Era, *available at* http://terrorism.about.com/od/types/a/Narcoterrorism.htm (discussing the melding of the war on terror and the war on drugs under the Bush Administration).

59. 50 U.S.C. § 1804(a)(7)(B).

60. Torin Monahan & Neal A. Palmer, *The Emerging Policies of DHS Fusion Centers*, 40 Security Dialogue 617, 625–30 (2009) (describing expansion of fusion center role and including a quote from a fusion center trainer: "If people knew what we were looking at, they'd throw a fit.").

discussion.[61] Professor Slobogin argues that a cabined exception is unneeded; under existing coerced confessions law, government interrogators have enough tools available. He also contends that the proposal would have negative consequences, predicting the inevitability of mission creep and "dangers"!

In this surrebuttal, I address these two sets of issues under the following headings: The Desirability of Providing More Guidance to FBI Interrogators and The Risk of Mission Creep.

The Desirability of Providing More Guidance to FBI Interrogators

Professor Slobogin argues that there is no need for the exception because interrogation methods of the type contemplated under the proposal already are legally permissible; that the only Supreme Court cases to the contrary are 50 years old.[62] He seems certain that these techniques would pass muster today and that the earlier cases are no longer good law. I believe, to the contrary, that confession law regarding deception, psychological stratagems, and the like is uncertain, and those earlier cases have not been overruled or repudiated.[63] True, there are also lower court cases that support his view. What he

---

61. Surprisingly, he also recharacterizes some features of the proposal as well as my positions on certain issues, and as a result, some of his contentions amount to a targeting of straw men. I do not plan to address such issues in this paper. Careful readers of my first paper and his reply can make their own judgments in the matter. For example, he suggests that I propose the use of "more strenuous" or "extraordinary" interrogation techniques, or that I contend that Supreme Court precedent supports the proposal. The actual proposal, in my view, is relatively modest, and I argue only that the exception *is not precluded* by Supreme Court precedent. The statement in *Quarles* cited by Professor Slobogin that seems to run contrary to the proposal regarded an issue not before the Court.

62. In the first paper (at footnote 24), I cited as examples only *Leyra v. Denno*, 347 U.S. 556 (1954) and *Spano v. New York*, 360 U.S. 315 (1959). Other similar cases include *Lynum v. Illinois*, 372 U.S. 528 (1963) and *Haynes v. Washington*, 373 U.S. 503 (1963); *also see* Rogers v. Richmond, 365 U.S. 534 (1961).

63. He cites as contrary more recent Supreme Court authority, but those cases do not directly address these issues and are not controlling authority.

does not acknowledge is that there are also lower court cases supportive of my position.[64]

At best, one can say that the state of law on this subject is uncertain, which creates a problem for government interrogators. The cabined exception proposal will help to relieve this uncertainty when it is most important to do so, in exigent circumstances[65]/terrorism investigations.

How does the FBI view the state of interrogation law? FBI policy, as set forth in the *Legal Handbook for Special Agents,* instructs that "no attempt be made to obtain a statement by force, threats, or promises," thereby setting up a reasonably bright-line standard of what is clearly prohibited. But the *Handbook* also states:

> Although it is not possible to predict in every case whether a court will find, under all the circumstances presented, that the statement was a product of the accused's free will or a product of coercion, there are predictable factors that a court will examine in making its determination.

---

64. *See, e.g.,* United States v. Anderson 929 F.2d 96 (2d Cir. 1991) (The agent "told Anderson three times to choose between having an attorney present during questioning or cooperating with the government. . . . these statements were false and/or misleading . . . . Agent Valentine's statements contributed to the already coercive atmosphere inherent in custodial interrogation and rendered Anderson's . . . confession involuntary as a matter of law"); United States v. Pichardo, 1992 WL 249964 (S.D.N.Y. 1992) (police officer's misleading statements about taking a lie detector test combined with the unfamiliarity of the defendant with the U.S. criminal justice system rendered the statement involuntary); United States v. Tingle, 658 F.2d 1332 (9th Cir. 1981) (The warnings that a lengthy prison term could be imposed, that Tingle had a lot at stake, that her cooperation would be communicated to the prosecutor, that her failure to cooperate would be similarly communicated, and that she might not see her two-year-old child for a while must be read together. . . . Viewed in that light, Sibley's statements were patently coercive.).

65. Professor Slobogin omits in most of his discussion the fact that my proposal is limited to exigent circumstances, and when he does acknowledge it at one point, he interprets exigent circumstances to refer to "ticking bomb" situations. In fact, contexts many degrees short of a ticking bomb fact pattern should qualify under an exigency element. *See, e.g.,* the standard set forth in the FBI's *Miranda* guidelines.

The *Handbook* then proceeds to list 11 factors, including:

Threats and psychological pressure; . . . Isolation, incommunicado interrogation; . . . Trickery, ruse, deception; . . . Promises of leniency or other inducements.

The statement of policy concludes with the following observation:

It must be kept in mind that the above factors are merely illustrative. The presence of any one or more of the factors mentioned above will not necessarily make a statement involuntary.

So what is an FBI agent to think? Can he or she, for example, use trickery, ruse, deceptions, other inducements, or psychological stratagems? The *Handbook*'s legal advice is, I believe, a respectable, albeit general and brief, description of the law in this area. Undoubtedly, however, it leaves agents uncertain about what they can do.

The *Handbook* approach also happens to suggest one type of structure that might be used for a cabined exception approach:

1. Establish a reasonably clear statement of what is prohibited under all circumstances.
2. Follow this statement with a list of interrogation practices which (given the uncertainty of the law) *may or may not* be permissible in interrogations involving ordinary crime or terrorism investigations where there are no indications of exigency.
3. Provide that in an exigent circumstances/threat-to-public-safety interrogation to obtain terrorism intelligence, any interrogation methods that do not violate the prohibited-under-all-circumstances standard (see 1. above) may be used.

Note: I am not suggesting that the specific FBI *Handbook* version of the always-prohibited standard nor the content of the *Handbook*'s totality-of-circumstances listing should be used; the *Handbook* suggests a type of structural approach that might be used. The exact terms in which the exception is cast remain to be determined, subject to the limitation that no extreme methods may be used under the exception.

The advantage of the cabined exception approach is not only that it will provide FBI interrogators with needed guidance and a safe harbor, i.e., more certainty,[66] about what they can do in exigent circumstance/threat-to-public-safety interrogations; it also will enable them to exercise greater flexibility in their choice of interrogation methods.[67]

The Risk of Mission Creep

Professor Slobogin and I share a concern about mission creep, that is, the risk that an exception, once established, might be extended, without express authorization, to other crime categories or extended in other ways. We differ, however, in our assessment of that risk. He cites the evolution of the Department of Defense's Total Information Program to illustrate the seeming inevitability of mission creep, and even uses it as the basis for a dramatic "give the government an inch" statement at the end of his paper. But mission creep in one government program does not mean that it will occur in another. A counterexample is the Foreign Intelligence Surveillance Act (FISA), legislated in 1978 as a broad exception to normal Fourth Amendment requirements. FISA, to be sure, can be criticized on various grounds, but for 33 years, operations under the statute have yet to be

---

66. That is, certainty compared with that under current law. Of course, no change in rules or structure can provide absolute certainty.

67. The recently promulgated FBI guidelines (barely mentioned by Professor Slobogin) relating to *Miranda* warnings in terrorism cases also are helpful: The exigent-circumstance/danger-to-public-safety formula used in the guidelines can serve as a model for the exigency element in the cabined exception proposal.

seriously faulted for mission creep; the FISA exceptional approach has not spread to ordinary crime areas.[68]

The FISA example illustrates how to control the risk of mission creep: Develop a principled basis for recognition of a special exception; establish clear standards in legislation or administrative regulations, or both; and subject the program to effective judicial review, meanwhile sensitizing all to the concern about the possibility of mission creep.

* * *

In conclusion, let me suggest another path to assess the merits of the cabined exception proposal. Professor Slobogin speaks of "dangers" posed by the proposal. Rather, the issue that should be considered is: What are the "dangers" if the course of action recommended by one or the other of us is followed but that individual turns out to be wrong in his assessment?

If I am wrong, an unnecessary exception will be created, and there will be a likelihood of mission creep, that is, the exception will spread to other crime categories and, overall, may compromise Constitutional standards governing coerced confessions.

But suppose Professor Slobogin is wrong? In that case, a cabined exception that could have been will not be created. As a result, some FBI agents, deterred by the uncertainty of the applicable coerced confession doctrine, may be reluctant to use particular interrogation methods and as a result fail to obtain information; and one can posit that in some instances that additional information might have enabled the Bureau to prevent a serious terrorist event.[69]

---

68. Professor Slobogin disagrees. He cites the legislative shift under FISA from a primary to "significant" purpose of the electronic surveillance as an example of mission creep. When Congress changes a rule, through the exercise of considered legislative judgment, I do not view that as "mission creep." Further, there is no evidence that this shift, which was legislated 10 years ago, has resulted in any significant invocation of FISA in ordinary crime investigations.

69. Additionally, because some other FBI agents use interrogation methods later determined to be improper, there may be more unsuccessful prosecutions of serious terrorists (than would otherwise have been the case).

So which is the greater set of dangers? Of course, the nature of the probabilities can affect judgment on the issue. And how does one weigh the risk of constitutional coerced confession doctrine being compromised against the chance that a serious terrorist event, possibly involving the loss of hundreds (or more) lives, might have been prevented? My judgment is that the latter is the graver risk, especially because I believe that the increase in the chances of obtaining relevant information and averting a terrorist event can be achieved without compromising basic societal values.

Exceptions: A Reply to Professor Abrams

Christopher Slobogin

Professor Abrams is to be commended for his efforts to reconcile confessions law with the threat posed by terrorism. After reading his two submissions in this debate, I am willing to contemplate an exception to *Miranda* where public safety is seriously threatened. But I cannot agree to Professor Abrams's exception to the prohibition on coerced confessions, which is the way he has framed his proposal. Here, I reiterate my concerns about his exception and set out a possible compromise position.

The Need for Clarity

Professor Abrams suggests (in his first note) that I mischaracterized his proposal by using words such as "strenuous" and "extraordinary" to describe the techniques he would permit. These words were only meant to indicate what he *might* be proposing (other than the citations to *Leyra* and *Spano*, he did not provide any guidance on that issue). I agree that *Leyra* and *Spano* do not involve "strenuous" techniques. But he now cites two other old Supreme Court cases that found a violation of due process (again without definitively declaring how they align with his exception): *Lynumn v. United States*,[70] where interrogators threatened to take away the suspect's children,

---

70. 372 U.S. 528 (1963).

and *Rogers v. Richmond,*[71] where interrogators threatened to arrest the suspect's wife. The techniques at issue in *Lynumn* and *Rogers* should not be legitimized even when terrorism is suspected, because they create too much pressure to say anything in order to avoid the threatened harm (as Lynumn herself explained[72]).

Professor Abrams argues that FBI agents need clarification as to the techniques they may use during interrogation. I agree, but an exception to the coercion prohibition is not needed for this purpose. Rather, current law, which admittedly is murky, should be elucidated. In the work that I mentioned earlier, I argued, in relevant part, that under current law a deceptive technique may be used during interrogation if: (1) "it is necessary (i.e. non-deceptive techniques have failed)" and (2) "it is not coercive (i.e., . . . would not be considered impermissibly coercive if true)."[73] Under factor (2), the question raised by the cases we have discussed is whether coercion would exist if the interrogator really were a friend (*Leyra*) or if, in the absence of a confession, the interrogator really would lose his job (*Spano*), the suspect really would lose her children (*Lynumn*), or the suspect's wife really would be arrested despite her innocence (*Rogers*). My brief answer (and probably the Supreme Court's answer as well) would be negative in the first two situations and, for the reason expressed above, affirmative in the last two.[74]

Professor Abrams acknowledges the potential for mission creep but does not think the latter techniques, if permitted when a national

---

71. 365 U.S. 534 (1961).

72. Lynumn stated, "I lied because the police told me they were going to send me to jail for 10 years and take my children, and I would never see them again; so I agreed to say whatever they wanted me to say." *Id.* at 532.

73. Christopher Slobogin, *Lying and Confessing*, 39 TEXAS TECH. L. REV. 1275, 1291 (2007).

74. For more detailed discussion, *see id.* at 1285–89. Although I am "certain" that the modern Court would endorse the "pretended friend" technique used in *Leyra*, I admit I'm not sure how it would come out in *Spano*. However, as I noted earlier, numerous lower courts have upheld similar tactics. Professor Abrams points out that some lower courts have been more punctilious about deceptive interrogation, but the fact that so many courts have not been indicates, at the least, that recognition of a special exception to the coerced-confession rule would be premature.

security exigency exists, would find their way into ordinary interrogations. Consider two reasons why he might be wrong. First, the notion of a "national security" investigation is extremely amorphous. I've already given some examples of this fact.[75] But consider one more. Shortly after 9/11, the FBI rounded up 4,800 Arab-Americans in Detroit, selected, according to Attorney General John Ashcroft, because "generic factors" suggested they might be terrorists; at the same time, around the country, another 15,000 other men and women, mostly of Middle-Eastern descent, were subject to FBI interviews.[76] If a cabined exception existed, it could easily apply here.

Professor Abrams might respond that his exception would not come into play unless "exigent circumstances" exist. But here the second mission creep problem arises. How does one define exigency? Professor Abrams himself states that he is willing to contemplate a concept "many degrees short" of the ticking bomb scenario. Although he probably does not mean to say so,[77] this latter language could easily be construed to mean that anyone the government thinks has intelligence about terrorism—in other words, anyone who might have fraternized with fundamentalists, traveled to Middle-Eastern countries, or visited suspect mosques—may be subject to coercive interrogation, a situation which is hardly "cabined."

---

75. Professor Abrams does not think that the recent amendment to the FISA statute is of major moment. But imagine a similar rule applied to interrogations of illegal immigrants. National security is not the *primary* concern in such cases, but it could easily be a *significant* one. Nor, contrary to Professor Abrams' suggestion, should congressional sanction of such a rule immunize it; the judiciary exists to ensure that Congress does not overstep constitutional boundaries.

76. *See* Christopher Slobogin, *Distinguished Lecture: Surveillance and the Constitution*, 55 Wayne St. L. Rev. 1124, 1125 (2009).

77. The confusion arises because Professor Abrams references the FBI's interpretation of the public safety exception as support for an expanded definition of exigency, but that statement consistently uses the words "immediate," "imminen,t" or "impending" in describing when interrogation restrictions might be relaxed, which is consistent with the ticking-bomb scenario. *See* FBI, *Custodial Interrogation for Public Safety and Intelligence-Gathering Purposes of Operational Terrorists Arrested Inside the United States*, Oct. 21, 2010, *available at* http://www.nytimes.com/2011/03/25/us/25miranda-text.html.

Professor Abrams raises the specter of a failure to prevent "a serious terrorist event" if his exception is not recognized. Again, there is little evidence agents need coercion to get this type of information. But if they do, coercive techniques should still be presumptively illegal; interrogators should be prosecutable or civilly liable for assaultive behavior, with the knowledge that they could assert a necessity defense. This is the approach taken in Israel, which has a much longer history of dealing with real terrorism than we do.[78]

An Exception to **Miranda?**

Professor Abrams makes several comments about the FBI's Guidelines and *Quarles*'s public safety exception, but they do not advance his argument appreciably. I did not discuss the FBI's approach in any detail because, in contrast to the *Army Field Manual,* it is noticeably lacking in specifics of the type just discussed.[79] And, as I noted before, *Quarles* declined to recognize a public safety exception to the coerced confessions rule, stating that Quarles was "certainly free on remand to argue that his confession was involuntary under traditional due process standards."[80]

However, the references to the FBI's Guidelines, which in part lay out when *Miranda* warnings are required, and to *Quarles,* which adopts a public safety limitation on *Miranda,* do suggest a more limited exception than the one proposed by Professor Abrams. Consider a rule declaring that when national security is threatened, interviewers need not inform suspects of their rights to silence and counsel, and can even state that there is no immediate right to cut off questioning or to obtain counsel. This rule would allow continuous questioning despite an invocation of rights, as long as the interrogation does not become coercive through means "calculated to break

---

78. Judgment of the Interrogation Methods Employed by the General Security Service, Israeli Supreme Court ¶¶ 35–36 (1999), *available at* http://www.derechos.org/human-rights/mena/doc/torture.html.

79. *See* LEGAL HANDBOOK FOR SPECIAL AGENTS, FACTORS AFFECTING VOLUNTARINESS, Section 7-2.2 (2003), *available at* http://fbiexpert.com/FBI_Manuals/Legal_Handbook_for_Special_Agents/FBI_Agents_Legal_Handbook.pdf .

80. 467 U.S. at 655 n.5.

the suspect's will."[81] Its constitutional justification would be that the Fifth Amendment only prohibits "compulsion" not the prophylactic rules created by *Miranda* and its progeny.[82]

Even this lesser exception may not be necessary, however. After all, the only remedy for a violation of *Miranda* (as opposed to a due process violation) is exclusion of a suspect's admissions from his or her criminal trial.[83] If the goal of the government is to prevent "a serious terrorist event" or to obtain intelligence against some third party, the government can ignore *Miranda* with impunity, because statements made during interrogation will only be used to stop the threat or to prosecute a person who lacks standing to exclude them. Even if the government wants to use the statements against the person interrogated, a simple *Miranda* violation would not require exclusion from a civil proceeding focused on preventive detention,[84] nor would it necessarily require exclusion from a military commission.[85]

Current interrogation rules do not stymie government efforts to sniff out and prevent terrorism. Adoption of an exception to the coerced confession rule would do real damage to suspects (most of whom will probably be innocent of any serious wrongdoing) and to the constitutional integrity of the American criminal justice system.

---

81. Oregon v. Elstad, 470 U.S. 298, 312 (1985). Based on a survey of false confession cases, Welsh White suggests a five-hour time limit for any single session. Welsh S. White, *Confessions in Capital Cases*, 2003 U. ILL. L. REV. 979, 1034.

82. Although *Dickerson v. United States*, 530 U.S. 428 (2000), held that *Miranda* is required by the Fifth Amendment, in no other decision since 1970 has the Court treated *Miranda* as anything other than a court-created rule subject to cost-benefit analysis. *Id.* at 450–61 (Scalia, J., dissenting).

83. Chavez v. Martinez, 538 U.S. 760 (2003).

84. Allen v. Illinois, 478 U.S. 364 (1986) (holding that the Fifth Amendment does not apply in sexual predator commitment hearings).

85. Hamdi v. Rumsfeld, 542 U.S. 507, 533–35 (2004) (holding that the "exigencies" of a military trial allow departure from normal procedures other than the "core elements" of "notice . . . and a fair opportunity to rebut the Government's factual assertions before a neutral decisionmaker").

Part Two: Data, Technology, and Privacy

Chapter Four
The Data Question:
Third-Party Information

Greg Nojeim

Orin Kerr

Introduction

I
f a suspected thief has left written records of his crime in a friend's desk, can the police simply subpoena the friend for the records in the desk or should that be treated as a search of the *suspect's* property?

That question is at the heart of the "third-party records doctrine," which has provided guidelines for criminal investigations since the late 1970s. In essence, the doctrine holds that information lawfully held by many third parties is treated differently from information held by the suspect himself. It can be obtained by subpoenaing the third party, by securing the third party's consent, or by any other means of legal discovery; the suspect has no role in the matter, and no search warrant is required.

Two well-known legal cases established the doctrine, *United States v. Miller*[1] (1976) and *Smith v. Maryland*[2] (1979).

In *Miller,* the defendant attempted to suppress evidence that investigators had obtained from his bank, arguing that he had an expectation of privacy under the Fourth Amendment. The Supreme Court held that because checks and deposit slips sent to banks are

---

1. 425 U.S. 435 (1976).
2. 442 U.S. 735 (1979).

freely circulated within the institution (the third party), Miller had no reasonable expectation of privacy and that law enforcement did not need a search warrant to obtain the data.

In *Smith,* Michael Smith had robbed Patricia McDonough and then phoned repeatedly to threaten her. The police secured a pen register at the phone company (third party) to trace the numbers of calls placed to McDonough. Smith appealed his conviction, asserting that the pen register had violated his Fourth Amendment rights. Justice Harold A. Blackmun wrote that when Smith voluntarily "conveyed numerical information to the phone company and . . . its equipment in the normal course of business, he assumed the risk that the company would reveal the information to the police."[3]

As more and more information moves online, some have questioned whether this principle should continue to be applied. For example, in the Global Positioning System (GPS) tracking case, *U.S. v. Jones*[4] (2012), Justice Sonia Sotomayor's concurrence described the third-party records doctrine as "ill suited to the digital age, in which people reveal a great deal of information about themselves to third parties in the course of carrying out mundane tasks." The principle remains the same: Suspects who entrusted their data to AT&T or Capital One in the 1970s are now entrusting their data to Google and Facebook. But the amount of data in the hands of third parties today is potentially much more revealing than in the 1970s. The question is whether that difference in quantity and quality has become a difference in kind.

Why the Third-Party Records Doctrine Should Be Revisited

Greg Nojeim

Under the third-party records doctrine, a person cannot assert a Fourth Amendment interest in information knowingly provided to a third party. If strict application of the doctrine ever served us well, it no

---

3. http://www.law.cornell.edu/supremecourt/text/442/735.
4. ___ U.S. ___, 2012 WL 171117 (S. Ct. Jan. 23, 2012).

longer does, leading to absurd results. This is particularly true in an age where so much more information is communicated through intermediaries. Proponents of the doctrine, quick to question what would limit it, fail to recognize that it already has holes that make its application uneven and illogical. It is time to turn the page and talk about limiting the third-party records doctrine to preserve the role of the Fourth Amendment in protecting personal privacy.

The doctrine was forged at a time very different from our own. There was no e-mail, and people communicated primarily by phone, fax, and letter. There was no World Wide Web; if you wanted to find merchandise, you used the Yellow Pages. Cellular telephones were the stuff of science fiction. To put the period in perspective, the Vietnam War was winding down, Americans heard their music on eight-track tapes, and the Chevy Nova and Ford Maverick were the leading car models.

Today, we live much of our lives online. We make friends, receive news, execute purchases, conduct business, create documents, store photographs, and find entertainment on the Internet. All these interactions create records in the hands of third parties about our interests, problems, loves and losses, finances, associates, family moments, and even our location at any moment. In days past, law enforcement could obtain this information only with an enormous expenditure of resources. It had to tail suspects and query informants. Consequently, the public could have some confidence that when law enforcement officers sought such personal information, they would have strong evidence of crime and focus their evidence collection on people likely tied to that crime. Fishing expeditions used to be expensive.

Today, the advance of technology has made it cheap and easy for law enforcement to gather this information secretly. The friction in the system and the practical anonymity that protected privacy are dissipating. And, with respect to law enforcement access, the third-party records doctrine has made it all possible.

The Supreme Court may not have been right when it decided *United States v. Miller* and *Smith v. Maryland.* Do people really "voluntarily" convey information to the phone company when they dial the numbers necessary to complete a call? And if numbers dialed on a telephone are voluntarily conveyed to the phone company, isn't the content of the call likewise voluntarily conveyed? Yet, call content has been Fourth Amendment–protected since *Katz v. United States* in 1967.

The third-party records doctrine rests on a false notion of consent: A person who surrenders information to a third party consents to that third party's disclosure of such information to others. In Fourth Amendment terms, this "expectation of privacy" is no longer "reasonable" because it is not reasonable to conclude that a third party will not disclose the information. The reality is quite different, though, almost akin to compelled consent, which is not consent at all. If you want to communicate efficiently today, your communications likely will go through your ISP's servers. The alternative means of communication involve either conveying information to other third parties or traveling to the other communicant so you can have a personal chat. Consent in this context has little meaning.

For too long, application of the third-party records doctrine has permitted absurd results. A person who stores documents and items in a physical space controlled by a third party in the business of renting it out retains a Fourth Amendment interest in those items.[5] But, if he or she stores the same information with an online lockbox in the business of providing online document storage services, he or she loses that Fourth Amendment protection, and it is available to law enforcement with a mere subpoena.[6]

Some worry that without the third-party records doctrine, clever criminals would convey evidence of their crimes to third parties so that those third parties would enjoy constitutional protection the criminals otherwise would lack. Criminals would substitute third-party services for activities that would otherwise be conducted in the open. But this proves too much: Innocent people would overwhelmingly constitute the group of "substitutors." The third-party records doc-

---

5. *See, e.g.,* Stoner v. California, 376 U.S. 483 (1964) (hotel room).
6. 18 U.S.C. 2703(b).

trine dissuades them from using the most desirable paths of communication by diminishing the privacy that would otherwise attend to them. For example, if the most efficient way for a group to edit a sensitive document is to store and edit it "in the cloud," why drive the group toward less-efficient solutions that don't involve an intermediary? That's what the third-party records doctrine does.

Some argue that the third-party records doctrine is simple for the courts to apply, and abandoning it or carving out exclusions would leave the courts at sea. But courts interpreting the Fourth Amendment have already set sail. They struggle to apply the "reasonable expectation of privacy" test and reach contradictory results when they do. Moreover, perhaps ease of application of a rule ought not to be the determinative factor in setting such rule when a fundamental right is at stake.

Where the third-party records doctrine would lead to undesirable results, Congress has sometimes stepped in with legislation to extend privacy protections to particular categories of data. For example, the Right to Financial Privacy Act (RFPA) protects financial records, the Video Privacy Protection Act (VPPA) protects video rental records, and the Health Insurance Portability and Accountability Act (HIPAA) protects medical records. However, each of these privacy statutes comes up short. They fail to extend the same protection that would come with full extension of Fourth Amendment protection to the records in question.

For example, the Electronic Communications Privacy Act (ECPA) requires a warrant based on a showing of probable cause for law enforcement access to the contents of electronic communications only for the first 180 days of the life of the communication. If a person's provider of electronic communications service holds the communications content for a lengthier period, the warrant requirement expires and law enforcement accesses the records with a mere subpoena.[7] The government, exploiting inartful language in the statute, takes the position that it can obtain opened e-mail from a pro-

---

7.   18 U.S.C. § 2703(a).

vider without getting a warrant no matter the age of the e-mail, leading to the absurd result that the spam e-mail you never open is better protected than the personal e-mail you do. Worse still, the ECPA provides that the information you outsource for storage and computing purposes is not subject to the warrant requirement at all, no matter its age and even though it is content.

The protections afforded by ECPA are no substitute for the warrant requirement that would attend full extension of Fourth Amendment protections to communications content in the hands of third parties. Like ECPA, the other privacy statutes that protect some categories of sensitive personal information generally do not require warrants for law enforcement access.

Professor Kerr agrees that full Fourth Amendment protection should be afforded communications content,[8] and he applauded when the Sixth Circuit reached that conclusion.[9] The court reasoned that e-mail today is analogous to a letter or a phone call, the contents of which are protected by the Fourth Amendment though each is conveyed through a third-party post office or telephone company.[10] So, there is consensus between us that at least one category of information—communications content—retains Fourth Amendment protection though conveyed through a third party. A chink in the armor of the third-party records doctrine.

Why, though, stop there? Is there no information, or collection of information, other than content that warrants Fourth Amendment protection when held by a third party? The *Smith* court may have thought so. It based its decision in part on the fact that numbers dialed on a phone and conveyed to the phone company to complete a call revealed little information—not the purpose of the communica-

---

8. Orin Kerr, *Sixth Circuit Rules that E-mail Protected by Fourth Amendment Warrant Requirement*, THE VOLOKH CONSPIRACY (Dec. 14, 2010) (http://volokh.com/2010/12/14/sixth-circuit-rules-that-e-mail-protected-by-the-fourth-amendment-warrant-requirement).

9. United States v. Warshak, 631 F.3d 266 (6th Cir. 2010).

10. United States v. Jacobsen, 466 U.S. 109, 114 (1984) and *Katz*, 389 U.S. 347, 353 (1967).

tion, not the identity of the parties communicating, and not even whether the call was completed.[11] This invites one to consider whether much more revealing non-content information would retain Fourth Amendment protection even if conveyed to a third party. While there might be a period of uncertainty while courts fill in the contours of the non-content information protected, that is no different from the way the courts have provided clarity to the application of other rights.

The Supreme Court's decision in *United States v. Jones*[12] may have shed some light. *Jones* did not involve third-party records—law enforcement agents tracked the defendant directly for 28 days by secretly attaching a GPS device to the vehicle he drove. The Court held that attaching a GPS device to collect location information was a Fourth Amendment search. While attaching the device was a "trespass" central to the Court's ruling, five justices signed concurring opinions that suggest that precise, pervasive monitoring of one's location could trigger Fourth Amendment protection without a trespass. This could bolster arguments that cell site-location information generated by use of a cellular phone is sufficiently sensitive to warrant Fourth Amendment protection though obtained from a third-party service provider. Two federal district courts have already reached that conclusion.[13]

Accumulations of location information held by cellular phone service providers might be one category of records to which Fourth Amendment protection should attach. What principles would limit the third-party records doctrine and determine the types of information that should retain Fourth Amendment protection though held by a third party? Scholars are already considering that question. Some say that one factor should be the relative sensitivity of the information, as the Court already hinted in *Smith*. Another might be whether

---

11. 442 U.S. at 741.

12. ___ U.S. ___, 2012 WL 171117 (S. Ct. Jan. 23, 2012).

13. *In re* Application of the U.S. for an Order Authorizing the Release of Historical Cell-Site Information (E.D.N.Y. 2011) (http://ia600309.us.archive.org/33/items/gov.uscourts.nyed.312774/gov.uscourts.nyed.312774.6.0.pdf) and *In re* Application of the U.S. for Historical Cell-Site Data (S.D. Tex. 2011) http://online.wsj.com/public/resources/documents/hughesorder1116.pdf (*appeal pending*).

the information is, in fact, knowingly "volunteered" to the third party. Still another might be whether the third party record holder has a relationship akin to a fiduciary relationship with the party to whom the record pertains.

The courts will ultimately determine the factors limiting the third-party records doctrine so that Fourth Amendment protection extends to information other than content. So that the Fourth Amendment retains its vitality in a networked society, it is important is that they start right away.

The Case for the Third-Party Doctrine

Orin Kerr

I'm delighted to debate Greg Nojeim on the controversial third-party doctrine of Fourth Amendment law. He and I share a number of first principles: Both of us are looking for a way to apply the Fourth Amendment to new technologies in a sensible and balanced way. Our disagreement is on how to do that. Nojeim would reject the third-party doctrine, while I would apply the doctrine in some cases but not in other cases. In this contribution, I want to explain why I think the much-maligned third-party doctrine is a critical tool for applying the Fourth Amendment to new technologies in some cases, but that it should not be extended to all cases.

My argument rests on the need to maintain the technological neutrality of Fourth Amendment protections. The use of third parties is akin to new technology, and that technology threatens to alter the balance of power struck by the Fourth Amendment. The third-party doctrine offers a way to maintain the balance of police power: It ensures that the same basic level of constitutional protection applies regardless of technology. Or so I will argue, drawing from two recent articles of mine: *The Case for the Third-Party Doctrine*, 107 Mich. L. Rev. 561 (2009), and *Applying the Fourth Amendment to the Internet: A General Approach,* 62 Stan. L. Rev. 1005 (2010).

My argument begins with a thought experiment. Let's start by imagining a world without third parties. If you want to send a package to a friend, you need to leave your home and carry it to your friend's house. If you want to go to the doctor's office, you need to visit it in person. In a world without third parties, you would need to venture out into the world on a regular basis to accomplish anything.

Next, ask yourself how the Fourth Amendment would apply to police investigations in this world with no third parties. The rules would be simple. The police would need a warrant to enter your home, but they would be permitted to watch you in public. They could watch you leave your home, travel to the home of your friend, and disappear inside when you delivered your package. They could watch you leave your home, go to the doctor's office, and disappear inside. All these steps would occur in public, where the Fourth Amendment offers no protection from the watchful eye of the police.

Now let's introduce third parties. Third parties allow individuals to do remotely what they would otherwise have to do in person. Instead of traveling to your friend to deliver the package, you can send the package through the postal mail. The postal network substitutes for your trip to your friend; instead of bringing your package to your friend, the mailman will do it. And instead of visiting the doctor in person, you could call the doctor on the phone. To borrow from the old advertising campaign for the Yellow Pages, you can "let your fingers do the walking." In each case, the third-party service means you no longer have to leave your home.

The critical point from the standpoint of Fourth Amendment law is that use of third parties introduces a substitution effect. In a world with no third parties, individuals often have to travel in public. The police can see when individuals leave their homes, where they travel, and when they arrive. Using third parties allows individuals to substitute a private transaction for that public transaction. Facts that used to be known from public surveillance are no longer so visible. By allowing individuals to use remote services, the use of third parties has brought outdoor activity indoors.

The question for Fourth Amendment law is how to respond to this technological shift. In my view, the goal should be to apply the Fourth Amendment in a technologically neutral way. In a world with no third parties, the Fourth Amendment strikes a balance of police power: It gives the government the power to investigate crimes in some ways, but also limits the government's investigations in important ways. I think that's a sensible balance, as it tries to balance our shared interests in deterring crime and punishing wrongdoing (which can only occur if the police successfully gather evidence to prove cases) with our commitments to privacy and avoiding government abuses of power. If I'm right that this balance is proper, then it follows we should maintain it. We should try to apply the Fourth Amendment so that it offers the same basic protections and strikes the same balance in a world of third parties than it did in a hypothetical world without them.

The third-party doctrine achieves that, in my view. The doctrine ensures that the Fourth Amendment applies to conduct that harnesses third parties in the same way it applies to events that occur without third-party help. It does so by matching the Fourth Amendment protection in the use of the third party with the Fourth Amendment protection that existed before.

Smith v. Maryland is a good example. In *Smith,* a robbery victim was receiving harassing phone calls. The police suspected Smith, and they asked the phone company to install a pen register on his home phone line. Whenever a call was placed from Smith's home phone, the phone company would record the numbers dialed and keep a record for the police. The record showed that the calls did indeed originate from Smith's home, and the police used that evidence to get a warrant to search the home and prove Smith's guilt. The question in the case was whether the numbers dialed were protected by the Fourth Amendment.

Before we get to the Court's reasoning in *Smith,* let's imagine what constitutional protection would apply if Smith could not use third parties but wanted to harass the victim anyway. Smith would have been forced to harass the victim in person: He would have left his house, walked to his car, and driven to her home. If the police suspected that Smith was the harasser, they could have watched him the entire way: All of Smith's conduct would have been exposed to public view without Fourth Amendment protection.

The Court in *Smith* ruled that Smith had no reasonable expectation of privacy in his numbers dialed. That rule maintained the level of Fourth Amendment protection regardless of whether Smith used a third party to harass his victim. By holding that the use of the pen register did not constitute a search, the Court ensured that the police would have the same information either way. The time of the call, the originating number of the call, and the destination of the call are the informational equivalents of what the police would have learned by watching Smith in public if he had not used a third party. In other words, Smith could not change the balance of Fourth Amendment protection by using a third party; the Fourth Amendment offered the same level of protection either way.

Importantly, my defense of the third-party doctrine implies an important limit: The doctrine should apply when the third party is a recipient of information, but it should not apply when the third party is merely a conduit for information intended for someone else. Put another way, the third-party doctrine should apply to the collection of non-content information in a network but not the contents of communications. The reason is that when the third party is merely a conduit for information, the information that is sent through the third party is not information that would have been revealed if no third parties had been used. In a world with no third parties, the message would remain private: If I bring you a sealed package in person, the government can't open up the package without a warrant. That same rule should and does apply if the delivered communication takes the form of a sealed letter in the postal mail, the contents of a phone call, or the contents of an e-mail.

Greg Nojeim makes several rejoinders to my approach. First, he argues that use of third parties is unavoidable in our modern world. That may be right, but the same was true about going out in public in the world without third parties. It would be unpersuasive to argue that the Fourth Amendment must protect what occurs in the public square because venturing out into the public square is unavoidable in our modern life. In my view, it is equally unpersuasive to claim that the Fourth Amendment must protect third-party substitutes because modern life requires their use.

Second, Nojeim argues that surveillance is cheaper and easier today than it used to be, and, therefore, that there is more of a need for legal regulation. Cost and time used to limit the government and channel its resources into the more important investigations and away from abuses; if the costs and time of surveillance drop, more abuses are likely unless the law steps up its role. This may be right. But Nojeim does not say why we should rely on the courts, rather than Congress, for this source of law. In my view, statutes are ideally situated to provide the kind of counterweight Nojeim envisions. Statutes have the flexibility to offer a tailored and nuanced approach to limit government properly without the blunt instrument of the Fourth Amendment and the warrant requirement.

Third, Nojeim suggests that it is arbitrary to apply the third-party doctrine to non-content information but then reject its application to the collection of contents. In his view, limiting the third-party doctrine to non-content information is "a chink in the armor" of the doctrine. But legal doctrines should apply to the extent of their rationale, and no further. Recognizing the validity of the third-party doctrine in some cases does not mean that it should logically apply to cases that are very different. In my view, countering the substitution effect of using third parties generates a sensible and reasonably clear rationale for both the presence of the third-party doctrine and its limits. To ensure the technological neutrality of the Fourth Amendment, we should apply the third-party doctrine where third parties generate substitution effects but not where they don't. This means that courts should apply the third-party doctrine to the collection of non-content information, but should reject the doctrine when the government collects contents of communications.

The Data Question: Reply to Orin Kerr

Greg Nojeim

As Orin Kerr points out, we do have some points of agreement. We both agree that the third-party doctrine should not apply to content held by third parties: A warrant should be required for content. Professor Kerr would stop there; I would not. I would not reject the third-party doctrine, but instead I urge more exceptions to its application. Content is not the only piece of information that should enjoy Fourth Amendment protection though held by a third party.

Professor Kerr does not contest my point that the Court's explanation for adopting the third-party doctrine in *Smith v. Maryland* rests on shaky ground. It reasoned that consumers *voluntarily* convey to the phone company the numbers they dial when placing a call, that they therefore have no reasonable expectation of privacy in that information, and have *consented* to further disclosure of this information to law enforcement. Perhaps we agree as well that "consent" in some scenarios in which the third-party doctrine applies is not really voluntary.

We also both agree that the Fourth Amendment should be applied to new technologies in a way that preserves the balance between privacy and law enforcement interests. I believe that the third-party doctrine has thrown that balance off. Technology permits us to communicate privately with people who are distant and numerous, but third parties often convey those communications. When they do, Fourth Amendment protections are lost.

Professor Kerr argues that requiring communicants to meet in person so law enforcement can observe them (but not listen in) maintains the balance by precluding a substitution effect. He worries that without the third-party doctrine obliterating Constitutional protection that might be afforded information about communications—the parties who communicated, the time of the communication, the length of the communication, the frequency with which they communicated, the location of the communicants, and the means by which the communication occurred, etc.—the bad guys would gain Fourth Amendment protection by using third parties instead of showing up in person.

This is not so. First, most communicants simply cannot meet in order to exchange information. It is not practical to do so, particularly where the communicants are far apart, or are numerous, as occurs much more frequently today than in the past. There is no substitution effect to worry about because there is no practical alternative to use of the third party to convey the information. The alternative is *not* to show up personally so the police can watch; rather, as a practical matter, the alternative when one wants to have a Fourth Amendment–protected communication is not to communicate at all.

Second, to the extent that there is a substitution effect, it falls primarily on people who are not involved in criminal activity because the overwhelming majority of communicants are not criminals. We should be more worried about people substituting less efficient means of communicating—showing up in person—in order to protect the privacy of information about their communications.

Instead of advancing technological neutrality, the third-party doctrine threatens it. The law is technologically neutral if the level of protection that information enjoys does not change when different platforms are used to convey the information. When the platform involves a third party, Fourth Amendment protection is often lost; but when the platform does not involve a third party, Fourth Amendment protection remains.

We agree that Congress ought to step in with statutory protections when application of the third-party doctrine leaves sensitive information unprotected from law enforcement access. But Congress generally legislates a weaker protection regime than the Fourth Amendment would require. For instance, Congress passed a law that is contrary to the rule to which we both agree: That the warrant requirement should apply to content held by third parties. Instead of requiring a warrant, ECPA allows law enforcement to use a subpoena when it seeks the contents of a document you have stored with a third party, the contents of your sent e-mail, and the contents of any message sent to you more than 180 days ago. There is no judicial authorization and no determination of probable cause, and notice that law enforcement has accessed this information is often delayed.

We both agree this should be changed. We may agree that other privacy statutes should also be strengthened. But I believe that court-authorized expansion of the list of exceptions to the third-party doctrine should also be part of the solution.

The Data Question: Reply to Greg Nojeim

Orin Kerr

Greg Nojeim and I have a relatively narrow disagreement: When applying the Fourth Amendment to a network, how far should the third-party doctrine extend to ensure that Fourth Amendment rules remain technologically neutral?

In my view, the third-party doctrine should not apply to contents of communications sent over networks but should apply to non-content information. Nojeim agrees with me that the third-party doctrine should not apply to contents. But he parts ways with me and argues that the third-party doctrine should apply to at least some non-content information. Nojeim does not specify which non-content information should fall within the doctrine, but he advocates a "court-authorized expansion of the list of exceptions to the third party doctrine" to cover at least some non-content information.

Our disagreement appears to hinge on a critical assumption. When we make the comparison across technologies, what starting point should be used? How much privacy do we enjoy on the baseline of a world without third parties, so that we can make a proper comparison to determine if a legal rule operates in a technologically neutral way?

In my view, history and tradition provide the proper baseline. Historically, when two people wished to communicate, they needed to do so in person. They needed to travel out in the open to meet somewhere. An observer wishing to watch them would be able to track their public movements. The observer would know when they walked to the meeting place, when they arrived, and when they left. To be sure, the outside observer would not know what the two people said to each other; the actual contents of the communication would be hidden from observation. But the information about the communication—who went where, and at what time, and for how long—could be observed in public.

Applying the third-party doctrine to non-content information but withholding it from content information re-creates this basic division for a networked world. It maintains the prior level of government power by ensuring that the contents of the communications remain protected and yet the non-content information about the communication—who went where, and at what time, and for how long—can still be observed. And it maintains that balance regardless of the particular network in question: The same balance applies if the network is the postal network, the phone network, or today's Internet.

When the third-party doctrine is appropriately limited to non-content information, the doctrine does not cause the imbalance that Nojeim suggests. The scope of privacy over communications networks remains the same as the scope of privacy in the physical world. It extends Fourth Amendment protection in some contexts and withholds in other contexts, just as does the Fourth Amendment when applied to physical space.

Chapter Five
National Security Letters

Michael German and Michelle Richardson

Valerie Caproni and Steven Siegel

Introduction

A National Security Letter (NSL) is a type of administrative subpoena used in terrorism and espionage investigations to obtain subscriber and transactional information from communications providers, financial institutions, and consumer credit agencies. Such information might include, for example, a suspect's name, address, place of employment, telecommunications toll records, financial data, and credit reports, depending on the type of NSL.

It would be legally and logistically impossible for the Intelligence Community (IC) to monitor the content of all communications transmitted by espionage and terrorism suspects. Instead, the IC frequently will focus first on "who is in contact with whom," rather than on what is being said or transmitted, to determine how to focus investigative resources appropriately. In the case of domestic communications, this approach may include the use of pen registers and trap-and-trace devices, which record the numbers being dialed from a telephone and the numbers of incoming calls (but not the actual content of communications), respectively. The "who is in contact with whom" rubric applies to the Internet as well; investigators may focus first on which IP addresses are in communication with each other rather than the specific content being transmitted. This focus extends to financial commerce, too, because it would be impossible to scrutinize all financial transactions of all terrorism and espionage subjects.

In cases of terrorism or espionage, the FBI utilizes NSLs to acquire basic information about its subjects. It may start with "who is in contact with whom" and, if appropriate and if the required showing can be made, then may transition to "what are they saying and doing?" Results from NSLs provided by communications providers can assist the FBI in identifying who is communicating with whom, but like pen registers and trap and trace devices, NSLs cannot yield the contents of communications.

Many federal agencies use administrative (non-court approved) subpoenas to obtain information relating to their duties; there are more than 300 instances in which the law grants such powers.[1] NSLs are a type of administrative subpoena that can be invoked *only* in terrorism and espionage investigations; they can be issued by the FBI to limited types of third-party records' custodians; the custodians are responsible for gathering and producing responsive materials to the FBI. The custodian can object if compliance would be burdensome, and the FBI cannot simply take materials from the custodian. For that reason, NSLs should not be confused with search warrants. Search warrants are issued based on a finding of probable cause by a neutral and detached magistrate; the person on whom a search warrant is served has no option to decline to cooperate and the entity serving the search warrant is authorized to seize material from the custodians. Moreover, the scope of a search warrant is set by the specific finding of the magistrate and can be quite broad, depending on the underlying facts. In contrast, NSLs have a strictly defined scope that has been set by Congress.

There are five types of NSLs. The first two were created by Congress in 1986—one under the Electronic Communications Privacy Act (ECPA) and another under the Right to Financial Privacy

---

1. Charles Doyle, *Administrative Subpoenas and National Security Letters in Criminal and Foreign Intelligence Investigations: Background and Proposed Adjustments*, Cong. Research Service (April 15, 2005), *available at* http://www.law.umaryland.edu/marshall/crsreports/crsdocuments/RL3288004152005.pdf.

Act (RFPA)—to assist FBI foreign intelligence investigations.[2] These statutes allow the FBI to obtain subscriber information, phone toll records and electronic communication transaction records (from a communications service provider), and financial information and transactions, such as the identity of the owner of a bank account and items deposited to and removed from an account (from a financial institution).

In the mid-1990s, Congress authorized a third type of NSL that can be issued under the Fair Credit Reporting Act (FCRA). Such records allow the FBI to discern which financial institutions are being used by a terrorism or espionage suspect. This information is necessary before an RFPA NSL can be issued to ensure the NSL is being served on the correct institution. To the extent that information is unknown, authority to issue an FCRA NSL facilitated that process by allowing the FBI to compel a consumer credit agency to identify the financial institutions with which a particular person had a relationship.[3]

Also in the mid-1990s, a fourth type of NSL was authorized via the National Security Act (NSA) for the specific purpose of investigating government employees with security clearance for leaks of classified information resulting in financial gain (i.e., espionage—this was a response to the case of Aldrich Ames).[4]

Finally, in 2001, a fifth type of NSL was created by the USA PATRIOT Act (USAPA), also under the FCRA. This second type of FCRA NSL, which can be used only in support of a counterterrorism investigation, requires a consumer credit reporting agency to furnish consumer reports and all other information in a consumer's file. This type of NSL is available to federal agencies authorized to conduct international terrorism investigations.

---

2. Charles Doyle, *National Security Letters in Foreign Intelligence Investigations: Legal Background and Recent Amendments*, Cong. Research Service (Sept. 8, 2009), *available at* http://www.fas.org/sgp/crs/intel/RL33320.pdf.

3. *Id.*

4. *Id.*

Changes to NSLs under the USA PATRIOT Act

The pre-USAPA certification requirement for *all* NSLs provided that an NSL could be issued only if it sought information "relevant to an authorized foreign counterintelligence investigation," and the requesting agency also certified that there existed "specific and articulable acts that the person or entity to whom the information sought pertained was a foreign power as defined by FISA." In 2001, the USAPA amended that standard to allow an NSL to be issued if the information sought from the recipient was "relevant to an investigation to protect against international terrorism or clandestine intelligence activities" (Section 505).

The USAPA also expanded the types of information that could be obtained via an ECPA NSL to include the form of payment used by the customer (Section 210). This change helped the FBI obtain information that could be used to confirm that particular communications were those of the person paying for the service, thus appropriately focusing FBI investigative resources. The USAPA also amended the Bank Secrecy Act to allow the Treasury Department to share financial information with intelligence agencies (Section 358). Finally, the USAPA forbade the issuance of an NSL in any investigation predicated solely on the exercise of First Amendment activities (Section 505).

By statute, all NSLs may carry a non-disclosure requirement, preventing the person or entity served with the NSL from telling others they had received it. For example, the ECPA statute read:

No wire or electronic communication service provider, or officer, or employee, or agent thereof, shall disclose to any person that the Federal Bureau of Investigation has sought or obtained access to information or records under this section.[5]

Although this so-called "gag order" had been in the statutes since they were first passed, subsequent to the USAPA they aroused con-

---

5. *Id.*

troversy. Following a series of legal challenges to the non-disclosure requirement, as part of the USAPA Reauthorization of 2005, Congress ultimately clarified the law to make clear that it did not preclude consultation with legal counsel; the changes also created a clear avenue of appeal for those served with an NSL.

In December of 2008, the U.S. Circuit Court for the Second Circuit held that imposing on the recipient of an NSL the obligation to commence a lawsuit to gain relief from the nondisclosure requirement was unconstitutional. The court made clear, however, that if the government was willing to take on that responsibility (i.e., the government would commence an action to enforce the non-disclosure requirement rather than the recipient commencing an action to obtain relief from the non-disclosure requirement), then the legislative regime would be constitutional. Although the FBI promptly modified its NSL practice to comply with the Second Circuit's decision, the statute has not been amended to conform to that decision regarding its constitutionality.

National Security Letter Statistics, 2004–2010[6]

Year	Number of NSL Applications	Number of Persons Involved in NSL Applications
2004	8943	N/A
2005	9475	3501
2006	12583	4790
2007	16804	4327
2008	24744	7225
2009	14788	6114
2010	24287	14212

6. "Foreign Intelligence Surveillance Act Court Orders: 1979-2010," Electronic Privacy Information Center (accessed Feb. 18, 2012), http://epic.org/privacy/wiretap/stats/fisa_stats.html.

Note: Statistics for the years 2003–2005 are incomplete; they don't include NSL requests that didn't identify whether the request was seeking information related to a U.S. person or a non-U.S. person. Also, the statistics do not include NSLs seeking only subscriber information.[7]

National Security Letters: The Need for Reform

Michael German and Michelle Richardson

The PATRIOT Act significantly broadened the authority of the Federal Bureau of Investigation (FBI) to obtain sensitive, private information about innocent Americans through national security letters (NSLs). This overbroad authority, combined with the FBI's disrespect for legal boundaries and its seeming inability to self-police, has resulted in the issuance of hundreds of thousands of NSLs, often targeting people two or three times removed from the subjects of investigations. The demonstrated abuse of privacy rights and civil liberties—and the absence of convincing evidence that NSLs are "indispensable tools" in the FBI's national security investigations—demand serious reconsideration of this authority.[8]

NSLs are secret demand letters issued without court approval or independent oversight to financial institutions, telecommunications and Internet service providers, and credit agencies to obtain sensitive personal information such as financial records, credit reports, the phone numbers and e-mail addresses with which a person has com-

---

7. The statistics do not include NSLs seeking subscriber information only due to the reporting format of the annual Attorney General's FISA Reports. *See, e.g.,* the 2010 FISA Report (http://www.fas.org/irp/agency/doj/fisa/2010rept.pdf): "In 2010, the FBI made 24,287 NSL requests (excluding requests for subscriber information only) for information concerning United States persons."

8. DEP'T OF JUSTICE, OFFICE OF INSPECTOR GENERAL, A REVIEW OF THE FBI'S USE OF NATIONAL SECURITY LETTERS: ASSESSMENT OF CORRECTIVE ACTIONS AND EXAMINATION OF NSL USAGE IN 2006, at 114 (Mar. 2008), *available at* http://www.usdoj.gov/oig/special/s0803b/final.pdf [hereinafter 2008 NSL Report].

municated, and possibly the websites a person visited.[9] The PATRIOT Act did not create NSLs, but before that act, only senior FBI officials could authorize their use, and the law required the FBI to certify that there were "specific and articulable facts giving reason to believe" the target of the NSL was an "agent of a foreign power." Section 505 of the PATRIOT Act removed both of these critical protections. First, and most problematically, it lowered the standard so that the FBI and other government agencies could obtain this sensitive data on the assertion that the information was merely "relevant" to an investigation, even if the person whose records were sought was not suspected of doing anything wrong. Second, it permitted NSLs to be issued by FBI field offices without review by high-level FBI officials.

Three Department of Justice Inspector General (DOJ/OIG) reports later confirmed pervasive FBI mismanagement and misuse and abuse of these PATRIOT Act–expanded authorities. And documents released pursuant to an American Civil Liberties Union (ACLU) Freedom of Information Act (FOIA) request revealed that the FBI also helped the Department of Defense (DOD) circumvent the restrictions Congress placed on its use of NSLs by issuing FBI NSLs for DOD investigations.

In 2007, the IG told the House Judiciary Committee that the FBI may have violated the law or government policies through the issuance of NSLs as many as 3,000 times since 2003, including as many as 600 "cases of serious misconduct."[10] An internal FBI NSL review conducted after the 2007 IG audit identified violations of law or intelligence policy that should have been reported to the President's Intelligence Oversight Board in 9.43 percent of the NSL files exam-

---

9. The four NSL authorizing statutes are the Electronic Communications Privacy Act, 18 U.S.C. § 2709 (2010); Right to Financial Privacy Act, 12 U.S.C. § 3405 (2010); Fair Credit Reporting Act, 15 U.S.C. § 1681 *et seq.* (2010); and National Security Act of 1947, 50 U.S.C. § 436(a)(1) (2010).

10. R. Jeffrey Smith, *FBI Violations May Number 3,000, Official Says*, WASH. POST, Mar. 21, 2007, *available at* http://www.washingtonpost.com/wp-dyn/content/article/2007/03/20/AR2007032000921.html.

ined, but the 2008 IG audit reexamined these files and found three times as many violations as the FBI did.[11]

The IG audits also confirmed that 40,000 to 50,000 NSLs were issued every year during the mid-2000s, and, in 2006, a majority of them were directed against U.S. persons.[12] This type of broad, suspicionless collection of private data about innocent Americans is the logical result of destroying the requirement of a factual nexus between an NSL and terrorist activity. And permitting the NSLs to be issued at the field office level removed the opportunity for centralized administrative oversight, making abuse more likely to occur, and less likely to be discovered by FBI managers.

With no internal controls and with complete disregard for the law, FBI agents soon ignored the minimal process involved in issuing NSLs and instead issued so-called "exigent letters," falsely claiming emergencies to obtain records without legal process.[13] These illegal requests—sometimes just a phone number written on a Post-it® note—were often given to the telecommunications companies with the promise that an NSL or grand jury subpoena would follow, but more often than not these promises went unfulfilled. Some agents found even Post-it notes too burdensome and instead asked company employees to just pull up a person's phone records so they could look over their shoulder to see whether a formal request such as an NSL was worthwhile.

There are also demonstrated problems with how the FBI handles data it receives in response to an NSL. Rather than using NSLs as investigative tools, as Congress clearly intended by only allowing them to be used when the information sought was relevant to an ongoing investigation, the FBI was using NSLs for mass data collection. The IG found FBI agents often carelessly uploaded information

---

11. *Id.*

12. Between 2003 and 2006, the FBI issued a total of 192,499 NSLs. In 2003, the FBI issued 39,346 NSLs; in 2004, it issued 56,507 NSLs; in 2005, it issued 47,221 NSLs; and in 2006, it issued 49,425 NSLs. 2008 NSL Report, *supra* note 1, at 110. In 2006, the last year for which complete NSL numbers are available, 57% of NSLs were issued to collect information on U.S. persons. *Id.* at 111.

13. 2008 NSL Report, *supra* note 1, at 86–97 (Mar. 2008).

produced in response to NSLs into FBI databases without reviewing it to evaluate its importance to the investigation or even to ensure the proper data was received. As a result, information received in error was improperly retained and illegally shared throughout the Intelligence Community.

The IG detailed several incidents where the FBI collected private information regarding innocent people not relevant to any authorized investigation, entered it into FBI case files, and/or uploaded it into FBI databases simply because the FBI agents requested records for the wrong phone numbers or for the wrong time periods. In two other incidents, information for individuals not relevant to FBI investigations was uploaded into FBI databases, *even though* the FBI case agent had written on the face of the documents: "Individual account records not relevant to this matter. New subscriber not related to subject. Don't upload."[14] Similarly, agents consistently failed to report or recognize when they received information from NSL recipients that was beyond the scope of the NSL request.[15] Agents self-reported the overproduction of unauthorized information in only four of the 557 instances the IG identified.

Congress foresaw some of these information-sharing and accuracy problems. In 2006, Congress voted to reauthorize other portions of the PATRIOT Act that were scheduled to expire. That legislation required the Attorney General and Director of National Intelligence to study whether minimization requirements were feasible in the context of NSLs. The report was due in February of 2007, and to date there is still no public information confirming that this report was ever sent to Congress, or even written. However, during the PATRIOT reauthorization efforts of 2009–2011, members of Congress did state that some type of internal minimization procedures were voluntarily adopted. Without public oversight, the effectiveness of these internal procedures in protecting the rights of innocent Americans remains in doubt. As the NSL saga reveals, internal controls unchecked by independent oversight are insufficient to prevent abuse.

---

14. *Id.* at 97 n.76.
15. *Id.* at 99 n. 1.

In addition to the overbroad scope of NSLs, there are constitutional problems with the non-disclosure or "gag orders" that accompany the overwhelming majority of NSLs. NSLs generally contain language prohibiting recipients from telling anyone besides a lawyer or the people necessary to comply with the NSL that they received it, much less what it requested. Because the letters go to the service provider, bank, or other third-party record holder, the target of the NSL—the individual whose records are sought or obtained—is never notified of the NSL or told that sensitive, personal information was disclosed.

The ACLU successfully challenged the constitutionality of the PATRIOT Act's original gag provisions, which imposed a categorical non-disclosure order on every NSL recipient.[16] In response, in 2006, Congress limited these gag orders to situations in which an FBI special agent in charge certifies that disclosure of the NSL request might result in danger to the national security, interference with an FBI investigation, or danger to any person.[17] Despite these revisions, the 2008 IG audit revealed that 97 percent of the NSLs issued by the FBI for the remainder of 2006 incorporated gag orders.[18] The ACLU challenged the gag order as rewritten and won again. The Second Circuit in *Doe v. Holder* held the gag unconstitutional because it put the burden on the recipient to prove that lifting the gag would not harm national security.[19] To be consistent with the First Amendment, the court shifted the onus to the government to demonstrate to a court a risk to national security whenever an NSL recipient notified the government that he or she wanted to challenge the gag. While the Obama administration testified before Congress that it was implementing its gag orders consistent with

---

16. *See* Doe v. Gonzales, 500 F. Supp. 2d 379 (S.D.N.Y. 2007); Doe v. Gonzales, 386 F. Supp. 2d 66 (D. Conn. 2005); Doe v. Ashcroft, 334 F. Supp. 2d 471 (S.D.N.Y. 2004); PIRA, Pub. L. No. 109-177, 120 Stat. 195 (2006); USA PATRIOT Act Additional Reauthorizing Amendments Act of 2006 (ARAA), Pub. L. No. 109-178, 120 Stat. 278 (2006).

17. Electronic Communications Privacy Act, 18 U.S.C. § 2709 (2006).

18. 2008 NSL Report, *supra* note 1, at 127.

19. John Doe, Inc. v. Mukasey, 549 F.3d 861, 884 (2d Cir. 2008).

this opinion, there is no public information to support this claim. The ACLU filed a Freedom of Information Act (FOIA) request to obtain more information.

The administration and Congress are not done with NSLs. In 2010, the Obama administration secretly requested that Congress expand its authority to collect a broad, undefined category of information called "electronic communication transactional records," which would allow the FBI to collect sensitive data, such as Internet use records, with NSLs. Despite debating the reauthorization of the PATRIOT Act off and on for two years from 2009 to 2011, the administration never once asked for this authority publicly, thereby preventing any meaningful debate about such a substantial expansion of authority.

Expanding the scope of NSLs is the last thing Congress should be considering as the executive branch's unilateral judgment of when and whether to gather this type of First Amendment–sensitive information is already suspect. The IG's 2008 audit included an episode in which the FBI applied to the Foreign Intelligence Surveillance Act (FISA) court for a Section 215 order,[20] only to be denied on First Amendment grounds. Section 215 orders are sought to obtain any tangible things relevant to a foreign intelligence investigation, including the records that can be obtained using an NSL. However, Section 215 orders require judicial approval and NSLs do not. In the cited example, the FISA court denied the FBI's request for this order twice, finding that "the facts were too 'thin' and [the] request implicated the target's First Amendment rights."[21] Rather than reevaluating the underlying investigation based on the court's constitutional

---

20. Section 215 of the PATRIOT Act allows the FBI to order any person or entity to turn over "any tangible things," so long as the FBI "specif[ies] that the order is 'for an authorized investigation . . . to protect against international terrorism or clandestine intelligence activities.'" Section 215 does not require the FBI to show probable cause or reasonable grounds to believe that the person whose records it seeks is engaged in criminal activity, or that the target is a foreign power or an agent of a foreign power.

21. DEP'T OF JUSTICE, OFFICE OF INSPECTOR GENERAL, A REVIEW OF THE FBI'S USE OF SECTION 215 ORDERS FOR BUSINESS RECORDS IN 2006, at 68 (March 2008), *available at* http://www.usdoj.gov/oig/special/s0803a/final/pdf.

concerns, the FBI circumvented the court's authority and continued the investigation anyway, using the broader unchecked authority provided in the NSL statutes in issuing three NSLs that were predicated on the same information contained in the unconstitutional Section 215 application.[22] We also know from one of the few unmasked NSL recipients that NSLs have been used to collect sensitive First Amendment activity in the past. Our client Doe—now publicly identified as Nick Merrill, the former operator of a small Internet service provider—believes that his NSL targeted someone because of his or her political speech on the Internet.

Ultimately, the NSL statute must be amended. While some of the management issues uncovered by Inspector General audits in the late 2000s may have been addressed, the fundamental problem remains the FBI's overbroad authority to obtain sensitive information relating to innocent people unilaterally without court review and without demonstrating any nexus to terrorism. This imprudently low standard remains an open door to abuse.

National Security Letters: The National Security Tool that Critics Love to Hate

Valerie Caproni and Steven Siegel

Congress first granted the Federal Bureau of Investigation (FBI) the authority to use National Security Letters (NSLs) in 1986. This authority ensured that the FBI would have the necessary tools to investigate threats to the national security posed by terrorists and spies because Congress had, at the same time, enacted statutory privacy protection to certain classes of records held by third-party businesses. Congress then understood a principle that remains true to this day: To appropriately and efficiently investigate threats to the national security, the FBI needs the ability to gather very basic information about individuals, including information about their finances, where they live and work, and with whom they are in contact, without alerting the targets that it is doing so.

---

22. *Id.* at 72.

The NSL authority is now and always has been quite limited. First, unlike grand jury subpoenas that can be used to collect any non-privileged document from any person or entity, the FBI can use NSLs only to obtain a very narrow range of information from a very narrow range of third-party businesses: NSLs can be used to obtain transactional information from wire or electronic communications service providers (e.g., phone companies and Internet service providers), financial institutions (e.g., banks and credit card issuers), and credit-reporting agencies. Other documents that can be critical to a national security investigation (e.g., hotel records, employment records) cannot be obtained using an NSL. Second, unlike grand jury subpoenas that can be issued in any type of criminal case, NSLs can be used only during duly authorized national security investigations. Finally, unlike grand jury subpoenas that can be issued by any Assistant U.S. Attorney or Department of Justice prosecutor, no matter how junior or inexperienced, NSLs can be issued only with very high-level FBI approval.

Although the NSL authority is quite limited, NSLs are nevertheless critical tools that enable FBI investigators to gather the type of basic information needed as the "building blocks" of national security investigations. It is not an exaggeration to say that virtually every significant national security investigation, whether of an individual suspected of planning to wreak havoc through an act of terrorism or of an individual suspected of spying on the United States for the benefit of a foreign nation, requires the use of NSLs for at least some critical information.

The arguments posited by ACLU's Mike German and Michelle Richardson are the ones that have been raised consistently by critics of NSLs, and must be considered against the backdrop of the importance of the tool and its limited scope. They assert that the standard required to issue an NSL is too low; the Department of Justice's Inspector General found that the FBI misused NSLs; and the so-called "gag order" is constitutionally objectionable. None of these arguments withstands scrutiny.

The Appropriate Standard

As noted in the German-Richardson essay, prior to the enactment of the PATRIOT Act, an NSL could be issued only if the FBI could certify that there were "specific and articulable facts giving reason to believe" that the target of the NSL (i.e., the person about whom information was sought) was an "agent of a foreign power." The PATRIOT Act changed the standard so that an NSL can now be issued so long as the information sought is "relevant" to an authorized national security investigation. Although the German-Richardson essay argues that this change in standard was problematic, it does so without providing any context. While reasonable people may disagree about the appropriate standard, a rational debate cannot occur in a vacuum.

First, the German-Richardson essay characterizes the information that can be obtained with an NSL as "sensitive, private information." A person not steeped in the intricacies of the law might infer from that assertion that an NSL can be used to obtain private diaries or psychiatric records or attorney-client privileged information—data that really is both sensitive and intensely private. The reality is far different. None of the information that can be obtained with an NSL is constitutionally protected. As noted above, the only information that can be obtained is information about who is in communication with whom (not the content of the communication); information contained in credit card and bank records; and information aggregated by credit-reporting agencies. The unifying feature of all that data is that it has been shared with a third party (e.g., the person on the other end of the phone line, the clerk in the store who processes a credit card purchase, the teller who processes the checks deposited in a bank account). As the Supreme Court made clear in *United States v. Miller,* 425 U.S. 435 (1976), there is no reasonable expectation of privacy in such data. Moreover, the overwhelming majority of NSLs are issued to obtain information that virtually no one considers "private" or sensitive: subscriber information for phone numbers.

Second, putting aside the hyperbole about the inherent sensitivity of the information, to consider whether the NSL standard is too low, one must consider whether the standard required in a national security investigation is in sync or out of sync with the standard that exists to get the exact same information in other contexts. The fact is that information obtainable with an NSL is also obtainable with a grand jury subpoena in any criminal investigation and with an administrative subpoena in narcotics investigations.[23] Although such investigations are obviously important, their purpose is to investigate crimes that generally pose far less danger to public safety and the national security than is posed by the targets of national security investigations. The standard for issuance of a grand jury or administrative subpoena is that the information sought must be relevant to the crime being investigated.[24] It would be exceedingly odd public policy to make it harder for investigators who are investigating threats to the national security to get basic transactional data than it is for investigators who are investigating routine federal crimes to get the exact same information.

The Inspector General Reports

German and Richardson assert that three Department of Justice Inspector General (DOJ/OIG) reports "confirmed pervasive FBI mismanagement and misuse and abuse" of the NSL authority. In fact, one IG report in 2007—almost five years ago—found significant weaknesses in the FBI's internal controls over the use of NSLs. Significantly, it did not find misuse in the sense of the FBI using NSLs maliciously or inappropriately to obtain records that were not relevant to an authorized FBI investigation. Indeed, the then-Inspector General testified that the IG "did not find that FBI agents sought to intentionally misuse the national security letters or sought information that they knew they were not entitled to obtain through the letters."[25] Instead, the IG found that in approximately 7.5 percent of

---

23. 21 U.S.C. § 876.

24. United States v. R. Enterprises, 498 U.S. 292, 299 (1991).

25. Statement of Glenn A. Fine before the Permanent Select Committee on Intelligence U.S. House of Representatives concerning "The FBI's Use of National Security Letters and Section 215 Requests for Business Records," March 28, 2007 (http://www.justice.gov/oig/testimony/0703b/final.pdf).

NSLs that it sampled (22 of 293), there was some type of error in either the issuance of the NSL or the handling of the data received. Of those errors, however, almost half (10 out of 22) were third-party errors, that is, the recipient of the NSL provided information that had not been sought by the FBI. Excluding third-party errors, the actual rate of FBI error was approximately 4 percent (12 out of 293). Of the 12 FBI errors, the overwhelming majority (10 out of 12) were non-substantive errors (e.g., the NSL used certification language slightly different from the statutory requirement (although the meaning was the same)) and only two (0.6 percent of all NSLs sampled) had substantive errors (i.e., one NSL sought information not appropriately obtainable with an NSL and one NSL was issued after the preliminary investigation to which it related had "lapsed"). While the FBI error rate was unacceptably high, one must question whether two substantive errors out of 293 NSLs can fairly be characterized as "pervasive."

The IG's second report on this topic was issued the following year, and it concluded that the FBI had made substantial strides in improving its processes and internal controls regarding the use of NSLs. Glenn Fine testified that the IG's "review of the FBI's corrective actions concluded that the FBI and the Department have evidenced a commitment to correcting the serious problems we found in our first NSL report and have made significant progress in addressing the need to improve compliance in the FBI's use of the NSLs."[26]

The IG's final report was not about NSLs at all but was about so-called "exigent letters." "Exigent letters" were devised and used primarily by a single unit at FBI headquarters to obtain phone records *without* issuing an NSL. While the practice originated in response to legitimate emergencies during which the FBI can obtain phone records without any legal process,[27] the practice morphed into an inappropri-

---

26. Statement of Glenn A. Fine before the House Committee on the Judiciary Subcommittee on the Constitution, Civil Rights, and Civil Liberties concerning "The FBI's Use of National Security Letters and Section 215 Requests for Business Records," April 15, 2008 (http://www.justice.gov/oig/testimony/t0804/final.pdf).

27. 18 U.S.C. § 2703.

ate substitute for required legal process (either a grand jury subpoena or an NSL) when there was no emergency.

While the German-Richardson essay focuses on the negative findings of the IG, it studiously avoids discussing the actual controls that are in place—and have been in place since shortly after the 2007 IG report—to avoid the type of errors discovered by the IG. Those controls are important because they collectively operate to ensure that the FBI is using the NSL authority responsibly and appropriately and in a way that is subject to audit and review.

First, by statute the FBI can use NSLs only in predicated national security investigations. The Attorney General has established the standards that must be met to commence a predicated national security investigation, and the predication for every full investigation of a U.S. person must be submitted to the Department of Justice for review.[28] The authority to conduct national security investigations is further controlled through internal FBI policy that establishes internal controls regarding opening predicated investigations.[29]

Next, since June of 2007, the FBI has had clear policies in place regarding virtually every aspect of issuing an NSL. NSLs may be issued by only a few high-ranking officials at FBI headquarters and by Special Agents in Charge (SAC) of FBI field offices.[30] The policy and procedures set forth clearly the standards that must be met before an NSL can be issued and mandate that the factual basis for the issuance of the NSL be documented in the file. FBI policy requires an FBI attorney to review every NSL before it may be authorized and clearly articulates the parameters for that review.

---

28. U.S. Dep't of Justice, Attorney General's Guidelines for Domestic FBI Operations (http://www.justice.gov/ag/readingroom/guidelines.pdf).

29. FBI, Domestic Investigations and Operations Guide (DIOG) (http://vault.fbi.gov/FBI%20Domestic%20Investigations%20and%20Operations%20Guide%20(DIOG)/fbi-domestic-investigations-and-operations-guide-diog-2011-version).

30. Special Agents in Charge are members of the Senior Executive Service. While there is no required period of time an agent would need to work at the FBI before he or she could be promoted to Special Agent in Charge of a field office, as a practical matter, most SACs have at least 15 years' experience before becoming an SAC.

Except in very limited circumstances, which generally account for fewer than 30 NSLs per year, NSLs must be created using an automated workflow system that minimizes the potential for error and helps ensure that statutory and policy requirements are met. As noted above, many of the errors detected by the IG in 2007 were non-substantive errors, such as citing the wrong statute, misquoting the required certification language, or omitting a step in the review process. The automated system ensures against such errors, automatically ensuring that each required review occurs and automatically ensuring the language in the NSL is uniform and legally correct. Finally, the automated system requires that documents received in response to an NSL are reviewed to minimize the impact of overproduction (production of material not called for by the NSL) and other third-party errors.

The FBI mandates that all employees who may play a role in issuing an NSL take training to ensure that they understand the rules. The FBI's Office of the General Counsel has created standardized training that is presented live and via a web-based training system. No person who is authorized to sign NSLs may do so until he or she has certified the receipt of that required training.

The FBI and DOJ have robust after-the-fact oversight to ensure compliance with the law, policies, and procedures. Attorneys from DOJ and FBI conduct audits of NSL usage in more than half of the FBI's field offices each year. Reports from these audits are presented to the Assistant Attorney General for National Security at DOJ, the FBI's General Counsel (GC), and the SAC of the field office that was audited, among others, so that any issues identified can be appropriately addressed. The FBI's Inspection Division also conducts an annual audit of NSL usage. The results of that audit are reported to the GC and the FBI's Deputy Director, among others. Employees who make certain errors that are detected during the Inspection Division process or through other means lose the authority to approve NSLs until they complete remedial training and attest that they understand the rules for NSL issuance.

In short, the FBI has taken numerous steps to improve compliance on the front end of the NSL process and to conduct rigorous self-evaluations after the fact to ensure strict compliance with the various statutes and policies that govern the use of this important tool.

Constitutionality of the Gag Order

By statute, the recipient of an NSL can challenge an NSL if responding would be unreasonable, oppressive, or otherwise unlawful. The FBI ensures that all recipients are aware of their ability to challenge the NSL by informing them of that right on the face of the NSL itself. Similarly, for any NSL that includes a non-disclosure order, the NSL notifies the recipient that, should they desire to disclose the fact that they received an NSL, they can either commence an action to set aside the nondisclosure requirement or they can notify the FBI of their desire to disclose. The NSL further informs the recipient that if the FBI wishes to maintain the secrecy of the NSL in the face of the recipient's desire to disclose it, the FBI will bear the burden of commencing a judicial proceeding in which it will be required to demonstrate the need for secrecy to a federal judge. If the FBI fails to do so, then the recipient will be free to disclose the NSL. That notification was added to all NSLs in February of 2009, based on the decision of the Second Circuit Court of Appeals in *Doe v. Mukasey,* 549 F.3d 861 (2d Cir. 2008), that such a procedure was necessary for the non-disclosure provision of the NSL statute to pass constitutional muster.

As the DOJ IG noted in its 2008 report on the FBI's use of NSLs, the vast majority of NSLs (approximately 97 percent) include non-disclosure requirements. That statistic is not particularly surprising, as NSLs can be used only to investigate national security cases—cases where the risk from premature disclosure can be particularly grave. For example, a terrorist target, on learning of an investigation, could take steps to expedite his plans of mayhem, to eliminate individuals who are believed to be cooperating with the government, or to destroy critical evidence. Or diplomatic relations could be gravely harmed if a foreign government were to learn that the FBI had obtained phone records associated with its officials who are in the United States. In short, there are good and sufficient reasons why the FBI generally wishes to keep the existence of an NSL secret; nevertheless, there is now a clear and constitutional process that NSL recipients can follow if they wish to make a disclosure.

We should note that after approximately three years of including in NSLs the provision for disclosure (during which time the FBI issued well over 50,000 NSLs), no recipients have notified the FBI that they wish to make a disclosure.

Conclusion

Although NSLs will no doubt remain the national security tool that critics love to hate, when one focuses on reality rather than on hyperbole, it is clear that the NSL is a constitutional tool that is reasonably used and is necessary in national security investigations to maintain the safety and security of the American people.

National Security Letters: Reply to the FBI

Michael German and Michelle Richardson

During the original PATRIOT Act debates, Attorney General John Ashcroft called librarians opposing the legislation "hysterical," and now Valerie Caproni and Steven Siegel argue that criticism of NSLs is "hyperbole." Caproni and Siegel repeat the FBI's previous assertions that NSLs are "critical tools" in the government's national security arsenal, but there is no public data to support this statement and, despite Caproni and Siegel's denials, there is ample evidence this overbroad authority has been abused, as any unchecked power usually is.

The Founders designed our constitutional system of government to prevent abuse of power through checks and balances between the branches and robust procedural protections where the government attempts to deprive an individual of his or her rights. Indeed, the most fully developed processes for the protection of civil rights exists within the criminal justice system, which makes the Caproni-Siegel comparison of NSLs to grand jury subpoenas most misplaced.

The grand jury, made up of ordinary citizens, is designed to serve as an independent check on law enforcement authority by protecting people against unfounded charges. As the *U.S. Attorneys' Manual* notes, the grand jury's power is limited by its narrow function of determining whether to bring an indictment for a criminal violation, which reduces the risk of unnecessary suspicionless data collection.[31] And in grand jury proceedings, the role of prosecutors, who are bound by the ethical obligations of their profession, is also a curb against law enforcement overreach. None of these protections exist with NSLs or other surveillance tools geared toward intelligence collection rather than criminal prosecution. The FBI has the sole discretion to issue NSLs with virtually no independent oversight. Moreover, a grand jury's indictment only starts the criminal justice process, after which additional rights attach and affirmative discovery obligations are imposed on the government. The government's obligation to disclose sources and methods of evidence gathering during trial is likewise a deterrent to improper collection, as the exclusionary rule compels suppression of illegally obtained evidence. The secrecy required in grand jury proceedings is designed to protect the privacy of the witnesses and individuals investigated, not to hide the government conduct from independent oversight and public accountability, as is the case with intelligence tools such as NSLs. Victims of NSL abuse have no way of knowing their rights have been violated, and no remedy.

The truth is that NSLs are intrusive tools. While the Supreme Court did fail to protect personal data held by third parties in 1976, as Caproni and Siegel point out, Congress then stepped in to protect financial, credit, and communications records, which most Americans consider sensitive and private information. The pre-PATRIOT NSL authorities Caproni and Siegel mention were limited to collect-

---

31. UNITED STATES ATTORNEYS' MANUAL 9–11.120, *available at* http://www.justice.gov/usao/eousa/foia_reading_room/usam/title9/11mcrm.htm#9-11.120.

ing information about suspected foreign agents or international terrorists. The PATRIOT Act expansion of NSL authorities allows the collection of data about any American the FBI deems "relevant" to an espionage or terrorism investigation, with no independent review. And given the technological advancements that have occurred since the Supreme Court's 1976 decision, which now leave vast amounts of personal information unprotected on third-party servers, trusting the government to be judicious with its access to such data through NSLs or other tools is even more misplaced.

Caproni and Siegel also note that NSL recipients rarely challenge the government's demands, which is not surprising given that NSLs seek records pertaining to someone other than the recipient. When the entities that hold private information show as little interest in protecting it as the government, everyone should worry. And it's interesting that in the three cases in which NSL recipients challenged these demands, the government withdrew the NSL requests rather than defend them in court, thereby mooting challenges to the underlying statute and throwing into doubt the government's justification for making these requests in the first place.

Finally, consider the FBI's continuing minimization of the abuse discovered by the Inspector General. The FBI's own audit found legal violations in 9.43 percent of its NSL files,[32] and the IG later determined that the FBI underreported the number of NSL violations by a factor of three.[33] These figures justify calling the abuse pervasive, and denying their importance only raises further skepticism that Americans can trust government agents with such unfettered power. The IG did indeed say the FBI made strides toward reform in 2008, but concluded, " . . . it is too soon to definitively state whether the new systems and controls developed by the FBI and the Department will eliminate fully the problems with the use of NSLs . . . ."[34]

---

32. 2008 NSL Report, at 8.
33. 2008 NSL Report, at 95.
34. 2008 NSL Report, at 15.

National Security Letters: Reply to the ACLU

Valerie Caproni and Steven Siegel

Even though National Security Letters (NSLs) cannot be used to obtain constitutionally protected information, Michael German and Michelle Richardson attempt to bolster their argument, or merely inflame the reader, against NSLs by casting the issue in constitutional terms. They imply that NSLs are used by the government to deprive citizens of their rights, even while admitting that the Supreme Court has ruled that the information that can be obtained with an NSL is not constitutionally protected. A critical reader may ask why they feel the need to couch the argument in this way.

The reality is that NSLs can be used only to obtain limited information that is held by third parties. In this regard, NSLs are a more limited tool than are grand jury subpoenas, which can be issued based on the same relevance standard but can be used to obtain a much broader range of information.[35] (NSLs are similar to administrative or grand jury subpoenas but can be used only to acquire specific categories of third-party records, such as phone toll records, credit reports, and bank records.) German and Richardson argue that the comparison to grand jury subpoenas is misplaced because the role of the grand jury and the ethical obligations of the Assistant U.S. Attorneys (AUSA) collectively serve as effective controls on the use of grand jury subpoenas; they argue that such checks do not exist for NSLs.

In practice, as they surely know, the grand jury itself plays little, if any, role in issuing subpoenas, and its function of determining whether to return a true bill of indictment follows—rather than precedes—the government's decision to issue grand jury subpoenas (to obtain third-party documents). The decision to issue a grand jury subpoena is typically made by a line prosecutor in conjunction with

---

35. *See, e.g.*, Permanent Provisions of the Patriot Act: Hearing Before the Subcomm. on Crime, Terrorism, and Homeland Security of the H. Comm. on the Judiciary, 112th Cong. 2 (March 30, 2011) (statement of Rep. Sensenbrenner, Member, House Comm. on the Judiciary, *available at* http://www.fas.org/irp/congress/2011_hr/patriot2.pdf).

a line law enforcement agent; involvement by higher-level prosecutors at the Department of Justice or higher-level agents at the investigating agency is the exception, not the rule. Moreover, there is virtually no after-the-fact review of the issuance of a grand jury subpoena by a prosecutor's supervisors or by DOJ to ensure that the information sought was, in fact, relevant to the underlying investigation and otherwise properly handled.

In contrast, contrary to the assertion by German and Richardson that NSLs are issued without effective oversight, all NSLs must be approved and signed by a high-ranking FBI employee (a Special Agent in Charge in a field office; Deputy Assistant Director or above at headquarters), and since 2007, all NSLs must also be approved by an FBI attorney who is bound by the same ethical obligations as AUSAs. In addition, there is significant after-the-fact oversight of NSLs by the Department of Justice's Inspector General and National Security Division that includes a review of the factual predicate for issuing an NSL and a review of how the responsive materials were handled. Finally, Congress receives semiannual and annual reports regarding the use of NSLs and has been briefed numerous times on their use.

For the past five years, since the DOJ Inspector General's 2007 report on NSLs, scrutiny of the use of NSLs has increased to include many of the controls discussed above. Although German and Richardson claim that that there is "ample evidence this overbroad authority has been abused, as any unchecked power usually is," they provide no evidence from recent years to support that assertion. They refer instead to the DOJ IG's 2008 report, which examined the FBI's use of NSLs in *2006*—six years ago and prior to substantial changes being made in the control environment for NSLs. Even then, they stretch the actual findings. They assert that the FBI audit found "legal violations in 9.43 percent of its NSL files." In fact, the FBI audits found *potential* violations in 9.43 percent of its files. Many of those potential violations involved third-party errors, not FBI errors. In any event, those findings are, five years later, largely irrelevant. FBI reviews subsequent to the changes adopted following the 2007

IG report have consistently found error rates below 1 percent. That progress is consistent with the DOJ IG 2008 report in which the IG acknowledged that the FBI and DOJ are committed to correcting the problems identified in the 2007 DOJ IG report and "have made significant progress in addressing the need to improve compliance in the FBI's use of NSLs."[36]

National security investigations are generally conducted in secret, and secrecy can spawn concern about unchecked power. There is always room for debate, and advocates such as German and Richardson serve an important role in ensuring that the public is aware of the tools being used in the national security arena and in articulating the risks associated with the use of such tools. The public debate surrounding the FBI's use of NSLs generated substantial changes in the FBI's internal processes and procedures and led to important changes in the law—changes that have enhanced privacy protections and have done so without hobbling the FBI's important national security work.

---

36. A REVIEW OF THE FBI'S USE OF NATIONAL SECURITY LETTERS: ASSESSMENT OF CORRECTIVE ACTIONS AND EXAMINATION OF NSL USAGE IN 2006, p.8 (March 2008), *available at* http://www.justice.gov/oig/special/s0803b/final.pdf.

Chapter Six
Einstein 3.0

Paul Rosenzweig

James X. Dempsey

Introduction

> If you can penetrate somebody else's networks remotely from the comfort of your office or an operating room, half a world away, in Xian . . . or somewhere in Siberia, or Tehran, maybe you don't need a spy. And the amount of information that you can exfiltrate that way is huge, so [cybersecurity developments have] changed the vector of the way of getting at information and it [has] turned the spy game [from] a retail into [a] wholesale operation, because you can take out . . . on a thumb drive now . . . more information than all the human spies in history have ever carried out of any country.[1]
>
> —Dr. Joel Brenner,
> former National Counterintelligence Executive,
> February 20, 2012

When warranted, we will respond to hostile acts in cyberspace as we would to any other threat to our country. All states possess an inherent right to self-defense, and we reserve the

---

1. Benjamin Wittes, Lawfare Podcast Episode #3: *Joel Brenner on America the Vulnerable*, LAWFARE, Feb. 20, 2012 (http://www.lawfareblog.com/2012/02/lawfare-podcast-episode-3-joel-brenner-on-america-the-vulnerable/).

right to use all necessary means—diplomatic, informational, military, and economic—to defend our Nation, our Allies, our partners, and our interests.[2]

—Department of Defense Cyberspace Policy
Report to Congress, November, 2011

Let me also be clear about what we will not do. Our pursuit of cybersecurity will not—I repeat, will not—include monitoring private-sector networks or Internet traffic.[3]

"Providing for the Common Defense": The Government as Internet Protector

Paul Rosenzweig

Einstein 2.0 is an intrusion detection cyber security system deployed by the federal government to protect the federal cyber networks. Its successor program, Einstein 3.0, not only detects cyber intrusions of the federal network but also actively seeks to prevent them. In the next iteration, these programs will likely be deployed on private networks to protect critical infrastructure. And therein hangs a tale—and a legal issue worthy of discussion.

A Short Introduction to Intrusion Detection and Prevention Systems

An intrusion detection system such as Einstein 2.0 operates through two principal systems. The first is what one might call a "look-up"

---

2. U.S. Dep't of Defense, Cyberspace Policy Report: A Report to Congress Pursuant to the National Defense Authorization Act for Fiscal Year 2011, Section 934 (November 2011) (http://www.defense.gov/home/features/2011/0411_cyberstrategy/docs/NDAA%20Section%20934%20Report_For%20webpage.pdf).

3. President Barack H. Obama, "Remarks by the President on Security of Our Nation's Cyber Intrastructure," *The White House*, May 29, 2009, *available at* http://www.whitehouse.gov/the-press-office/remarks-president-securing-our-nations-cyber-infrastructure.

system. Every piece of malicious code is unique—it has what is known as a "signature" (essentially, an identifying code component that serves as a marker for the program). The detection program has a database of known malicious code signatures on file and constantly compares incoming messages with the malicious signatures. When it finds a match, it sends an alert to the recipient. For the federal system, Einstein 2.0 gets its database of malicious signatures from a variety of sources, including both commercial sources such as Symantec (a private Internet security company) and classified sources at the National Security Agency (NSA).

The second system, less definitive and more probabilistic, utilizes what is known as "anomaly detection." In essence, the Einstein 2.0 program knows what "normal Internet traffic" looks like and can produce an alert when the incoming traffic differs from normal by some set tolerance level. Notably, the Einstein 2.0 system is a gateway system that screens traffic as it arrives at federal portals and does not stop any traffic.

Einstein 3.0, the next generation of the program, is based on a classified NSA program known as Tutelage and is different in several respects. Its goal is to go beyond detection of malware (and an alert) to actual intrusion prevention. After all, simple detection is a bit like telling someone "you're being robbed" after the bank robber is already inside the vault. It is, naturally, far more valuable to prevent the robber from getting into the bank in the first instance. To do this, Einstein 3.0 must intercept all Internet traffic bound for federal computers before it is delivered, delay it temporarily for screening, and then pass it along or quarantine the malware as appropriate. And for that system to be effective, the Einstein 3.0 screening protocols must reside *outside* the federal government firewalls, on the servers of trusted Internet connections.

The Scope of the Problem

In considering the utility of any proposed intrusion detection or prevention system, it is useful to have some idea of the scope of the problem. Clearly, if the problem is of modest scope, we would be less likely to authorize activities that posed a risk to privacy and civil liberties. Contrarily, the more significant the problem, the more likely we are to push our defensive measures to the outer limits of law.

To my mind, there can be little doubt that the challenges of securing the web are significant. At a macro level, the absolute volume of the malfeasant activity on the Internet sometimes defies human understanding. One study of Internet traffic on the Bell Canada network found that more than 53 gigabytes per second (!) contained malicious code of some type (this is roughly 17 percent of the total traffic load at any one time).[4] Likewise, the security firm Symantec discovered 286 million new unique malicious threats in 2010, or roughly 9 new malware creations every second.[5] As Noah Shachtman recently stated, even putting aside the more existential threats from cyber war or cyber espionage, the growth of crime on the Internet is quickly making the network a bit like the South Bronx in the late 1970s—a place that no reasonably sensible person would want to go.[6] No surprise then that estimates of losses to cyber crime are pretty significant. For example, the consulting firm Detica has estimated the annual loss from cyber intrusions in the United Kingdom at £27 billion.[7] Two years earlier, McAfee Security estimated annual cyber-crime losses at $1 trillion globally.[8]

4. *Combating Robot Networks and Their Controllers* (Unclassified Version 2.0, May 6, 2010), http://www.scribd.com/doc/51938416/Botnet-Analysis-Report-Final-Unclassified-v2-0. One of the authors of the report, Rafal Rohozinski, gave a colloquial talk on this study to the St. Galen Symposium earlier this year. *See* http://www.youtube.com/watch?v=DpRYXRNWka0&feature=youtu.be.

5. Christopher Drew & Verne G. Kopytoff, *Deploying New Tools to Stop the Hackers*, N.Y. TIMES, June 17, 2011, sec. Technology, *available at* http://www.nytimes.com/2011/06/18/technology/18security.html?_r=1&scp=1&sq=hackers%20symantec &st=cse.

6. Noah Shachtman, *A Crime Wave in Cyberspace*, WASH. POST, July 22, 2011, *available at* http://www.washingtonpost.com/opinions/a-crime-wave-in-cyberspace/2011/07/21/gIQAYfbIUI_story.html.

7. *The Cost of Cyber Crime,* DETICA (Feb. 14, 2011), *available at* http://www.detica.com/uploads/press_releases/THE_COST_OF_CYBER_CRIME_SUMMARY_FINAL_14_February_2011.pdf.

8. Elinor Mills, *Study: Cybercrime costs firms $1 trillion globally*, CNET NEWS (Jan. 28, 2009), http://news.cnet.com/8301-1009_3-10152246-83.html.

At a more micro level, we recently were reminded of the remarkable extent of even a single intrusion, when McAfee released a report on a systematic intrusion it dubbed Operation Shady RAT.[9] There they detailed a systematic, five-year intrusion that infected more than 70 different victims for periods of time ranging from less than a month to more than two and a half years. Given the sophistication of the attack and the apparently non-economic nature of the targets (several, for example, were Olympic-related and were compromised just prior to and during the 2008 Olympics in Beijing), McAfee speculated (correctly, in my judgment) that the program was the work of a nation-state actor. If so, it joins Operation Aurora (the Chinese-sponsored attack on Google), GhostNet (an infiltration of the Dali Lama's computer system), and Operation Night Dragon (a systematic infiltration of oil and gas companies) in a growing list of significant compromises that have begun to take on national security dimensions.

The Legal Debate

There is little real debate over the operation of Einstein 3.0 as applied to government networks. I have every reason to expect that my colleague in this debate and I both agree that it is appropriate and necessary for the government to monitor traffic to and from its own computers. Rather, our disagreement is likely to be over how deeply a government-owned and government-operated system such as Einstein (call it Einstein 4.0, if you want) may be inserted into private networks, either to protect the government or to protect private-sector users.

In my view, it is likely that such a system would pass constitutional muster, though its full operation would require the amendment of several existing statutory restrictions—amendments whose political viability is highly questionable. [An important note: I discuss here the legality of the system, not its advisability as a matter of policy.]

---

9. Dimitri Alperovich, *Revealed: Operation Shady RAT* (McAfee, July 2011), http://www.mcafee.com/us/resources/white-papers/wp-operation-shady-rat.pdf.

Content v. Non-content

To begin with, current doctrine makes it clear that the non-content portions of the intercepted traffic are not protected by the Fourth Amendment. The Supreme Court addressed these questions in a related context in two 1970-era cases: *United States v. Miller*[10] and *Smith v. Maryland.*[11] In both cases, the question was, in effect, to what degree did an individual have a constitutional protection against the wholesale disclosure of information about him that had been collected by third parties? And in particular, could an individual use the Fourth Amendment to prevent the government from using data it had received from a third-party collector without a warrant?

The Supreme Court sided with government, developing what has come to be known as the "third-party doctrine." In *Miller,* the Court held that financial information voluntarily disclosed by an individual to a bank was not protected by the Fourth Amendment against subsequent disclosure to the government. Likewise, in *Smith,* the Court held that an individual's toll records—records of the phone numbers called by the individual—were not protected against disclosure. It seems to follow almost a fortiori that non-content header information in Internet traffic (IP addresses, "to" and "from" lines, and such) are likewise not protected as a matter of constitutional law.

Consent and Banners

The *Miller* analysis will not, of course, permit the use of an intrusion prevention system to routinely scan the content portions of an Internet exchange. But those may also be the portions of a message that contain malware. Indeed, it would be an extremely poor rule that permitted screening of only non-content information for malware, as that would simply draw a map for malfeasant actors about how to avoid the intrusion detection systems.

For traffic directed to federal computers, the content/non-content distinction is comparatively easy to solve. As the Department

---

10. 425 U.S. 435 (1976).
11. 442 U.S. 735 (1979).

of Justice (DOJ) assessed it, Fourth Amendment concerns (and also most statutory concerns) can be addressed by using a robust form of consent.

Interestingly, the consent concerns are more for the recipient (some federal employee or agency) than for the sender. Not unreasonably, DOJ concluded that the sender loses his privacy interest in the content of an Internet communication when it is delivered.[12] After all, he intended the recipient to get the message and lawfully the recipient may do with it what he likes (including putting it in the spam folder). There is little, or no, interest (once the mail is delivered to its intended recipient) in fooling the recipient into doing the wrong thing.

So, in Einstein 2.0 and 3.0, the main consent concern is actually for the recipient employee, who might have a privacy interest in the contents of the e-mail, as against his employer who would be screening the content of his incoming mail.[13] As to those employees, however, the government can (and in my experience does) make consent to e-mail monitoring a condition of employment, reinforced by log-in click through banners that warn the employee that his e-mail will be monitored. In the past it has done this, lawfully, to prevent government resources from being used for illegal or inappropriate purposes (for example, downloading pornography), and that legal meme has simply been translated to the cyber security realm.

This analysis paints a good road map for how the government can (and has begun to) expand its presence into the private sector (where neither the sender nor the recipient is a federal employee or agency). The extension has begun with voluntary agreements with closely aligned government contractors in the defense industrial base.

---

12. Legality of Intrusion Detection System to Protect Unclassified Computer Networks in the Executive Branch (DOJ, Office of Legal Counsel, August 2009), http://www.justice.gov/olc/2009/legality-of-e2.pdf; Legal Issues Relating to the Testing, Use, and Deployment of an Intrusion Detection System (Einstein 2.0) to Protect Unclassified Computer Networks in the Executive Branch (DOJ, Office of Legal Counsel, Jan. 2009), http://www.justice.gov/olc/2009/e2-issues.pdf.

13. O'Connor v. Ortega, 480 U.S. 709 (1987) (government employee has privacy expectation in the contents of his desk at work).

To foster their ability to do business with the federal government, those companies agree to monitor incoming Internet traffic using government-provided threat signature information. Here, again, as in the case of communications bound for the federal government, the non-content-addressing information is not protected by the Fourth Amendment, the senders have no expectation of privacy as against the recipient, and the recipient employees consent to scrutiny of the communications as a condition of employment.

This "voluntary" consent model is readily expandable to almost any industry that depends on federal financing. Already, there is talk of extending this model to the financial and nuclear industries.[14] A more problematic extension might be to the health-care industry or the education community, but those problems are likely more ones of policy than of law. However broad the expanse of this voluntary consent model, it seems likely to occupy a significant fraction of the field.

But what, then, about the remaining Internet traffic—private-to-private traffic that is not directed to or from a critical infrastructure industry? Here, the legal limitations on the scrutiny of private content network traffic are at their highest and, in my judgment, are likely to prevail. But let me offer two theoretical grounds on which the government might proceed (while hastening to add that I would not approve of these as a policy matter).

First, one might argue that wholesale scrutiny of network traffic is reasonable based on the prevalence of malware in Internet traffic. If the government were to minimize non-malware traffic intercepts and eschew criminal prosecution, it could argue that broad-based scrutiny is akin to a sobriety checkpoint on the highway, a necessary special-needs administrative inspection that is acceptable precisely because of the harm it averts.[15]

---

14. David Ignatius, *Department of Internet Defense*, WASH. POST (Aug. 12, 2011), *available at* http://www.washingtonpost.com/opinions/department-of-internet-defense/2011/08/12/gIQAPQcxBJ_story.html.

15. Michigan Dept. of State Police v. Sitz, 496 U.S. 444 (1990). The limitation to non-criminal sanctions is likely a critical distinction that bears on the constitutionality of any program; City of Indianapolis v. Edmond, 531 U.S. 32 (2000) (disapproving checkpoint drug searches).

Somewhat more ambitiously, the government could adopt a law making consent to malware intrusion detection systems an implied condition of access to the Internet. Here, the analogy would be to the implied consent laws that have been adopted by many states with respect to sobriety tests. In these states, acceptance of a driver's license and the right to travel on the public roads brings with it mandatory consent to a sobriety test—refusal itself is a crime.[16] One can imagine the adoption of such a regulatory system for the Internet.

All of which is to say, by way of introduction to the topic, that there is probably a wide scope of constitutionally permissible activity for the government, even on the private networks of the Internet. Exactly how far the government is permitted to go, in the end, is more likely a question of wise policy than it is of constitutional law.

Liberty and Security Weigh in Favor of Private Sector Leadership

James X. Dempsey

The question here is stated simply enough: Between the federal government and the private sector, who should have the lead in conducting cybersecurity monitoring of the privately owned and operated communications infrastructure and of private-to-private communications? Paul Rosenzweig and I both agree that the federal government properly has broad authority to monitor its own networks for cybersecurity purposes. We also agree that the private sector properly has broad authority to monitor its systems.

Our point of divergence is over whether the federal government can do a better job of monitoring the private sector networks and private-to-private communications passing over them than the private sector owners and operators of those networks could themselves do. In my view, the private sector is both more agile and more knowledge-

16. For example, an Illinois statute (625 ILL. COMP. STAT. § 5/11501.1, ILL. REV. STAT. ch. 95 1/2, 11-501.1 (1981)) provides that any person who drives an automobile in that state consents to take a breath-analysis test when requested to do so by an officer as incident to an arrest for driving while intoxicated. Other states have similar laws.

able in key respects about its systems than the federal government could ever be. To the extent that the federal government has some specialized knowledge that would be helpful to the private sector, the goal of policy should be to transfer that knowledge to the private sector in a way that is both secure and useful. Leaving the main responsibility for protecting private sector networks in the hands of the private sector will not only be most effective from a security standpoint, but it will also have significant civil liberties benefits as well.

Don't Get So Scared That Effectiveness Is Ignored

Paul Rosenzweig has done such a good job of describing the variants of Einstein that I have to add only a few caveats. And he is correct in emphasizing that just because something is constitutionally permitted and otherwise legal does not mean that it would be wise or effective policy. Too often since 9/11, the question of a program's desirability has been reduced to the question of whether it is lawful. Rosenzweig has moved past that narrow question, and any Administration should do likewise. However, on the path to his policy analysis, Rosenzweig misframes a few significant points and fundamentally misses what should be a major step in any analysis.

First, in responding to the challenges of cybersecurity, it is best to appreciate that there is some quantum of hyperbole in descriptions of the problem. Estimated costs of cybersecurity losses are almost certainly unreliable and most likely exaggerated.[17] Moreover, there are policy risks that come from overstated claims of weakness.[18] The sense of urgency—the need to do something—can crowd out a more balanced analysis and overshadow the question of efficacy. That is precisely what seems to have happened in Paul Rosenzweig's analysis. He jumps from his statements of acute threat to his assumption

---

17. *See* DINEI FLORENCIO & CORMAC HERLEY, SEX, LIES AND CYBER-CRIME SURVEYS, Microsoft Research, *available at* http://research.microsoft.com/pubs/149886/SexLiesandCybercrimeSurveys.pdf.

18. Jerry Brito and Tate Watkins, *Loving the Cyber Bomb? The Dangers of Threat Inflation in Cybersecurity Policy*, Mercatus Working Paper (April 2011), *available at* http://mercatus.org/sites/default/files/publication/WP1124_Loving_cyber_bomb.pdf.

that government intervention into private networks is necessary. This overlooks the fact that the private sector, and particularly the network operators, have made huge investments in cybersecurity.[19] The marketplace has seen the development of many products, services, and practices for greater cybersecurity, and these have been effective in preventing, mitigating, or responding to a range of vulnerabilities and threats.[20] At the same time, of course, it must be admitted that the federal government, even the defense and intelligence agencies, have not done a good job of getting their own cybersecurity houses in order.[21]

Surely, any reasonable observer would still conclude that the cybersecurity problem is very serious and that more needs to be done. Objectively, however, it is impossible to say that the federal government is better positioned to solve the problem than the private sector. With a more cautious description of threats and a deeper appreciation of responses to date, the formulation of solutions could proceed in a more balanced fashion.

Second, in considering the differences between signature-based defensive measures and more probabilistic approaches, it is important to recognize that both methods can produce false positives as well as false negatives.[22] This is important because any sophisticated cybersecurity activity at the network level today will include an element of human intervention, judgment, and response. At the net-

---

19. Greg Nojeim, *Cybersecurity: The Power of Private Sector Solutions* (Jan. 15, 2010), at http://www.cdt.org/blogs/greg-nojeim/cybersecurity-power-private-sector-solutions.

20. *Improving our Nation's Cybersecurity through the Public-Private Partnership* (March 2011) at p. 5, at http://www.cdt.org/files/pdfs/20110308_cbyersec_paper.pdf.

21. A steady stream of GAO reports documents the limitations of the federal government's own response. *See*, recently, *Information Security: Weaknesses Continue Amid New Federal Efforts to Implement Requirements* (October 2011), at http://www.cdt.org/files/pdfs/20110308_cbyersec_paper.pdf.

22. *See* Lawrence Walsh, *McAfee Antivirus Snafu Crashes Windows XP Machines* (2010) (signature file update falsely identifies valid Windows process as malicious), at http://www.channelinsider.com/c/a/Security/McAfee-Antivirus-Snafu-Crashes-Windows-XP-Machines-568703/.

work level, cybersecurity is not conducted in a black box. Actual humans look at actual communications. This is directly relevant to the twin questions we are facing in this debate: Who is in the better position, from a purely operational perspective, to conduct that human intervention: the private-sector owners and operators of the infrastructure or government agents who do not fully understand the network? And, second, from a civil liberties perspective, what are the risks of passing large quantities of data to the federal government? On both fronts, I believe, the private sector is better positioned to monitor and protect private-to-private communications.

Our legal system for communications regulation is based on the privatized management of our communications networks. Among other benefits, this has placed a buffer between private communications and the government. That buffer seems one worth maintaining, while the task of creating an oversight system of checks and balances to replace it seems entirely beyond our current political and administrative capabilities. (The administration's cybersecurity legislation proposes the creation of a government cybersecurity analysis center that does not yet exist to be operated under rules yet to be written and overseen by a body that is still without members or staff more than four years after it was authorized by Congress.)

The analysis of those urging greater government intrusion into private networks repeats a fundamental flaw of national security analysis since 9/11: The fact that a problem is very serious tells us nothing about the best way to address the problem. It is still necessary to consider the effectiveness of any proposed solution. When it comes to cybersecurity, the government is in the best position to take the lead on monitoring government systems and the private sector in the best position to monitor private networks and communications.

The Legal Issues

I agree with Paul Rosenzweig that the policy considerations (including the question of efficacy) should probably be dispositive in weighing deployment of Einstein to monitor private-to-private communications, but I disagree with his minimization of the legal concerns. In particular, I do not think that Einstein can be given a pass under the claim that non-content is not constitutionally protected. The distinction between content and non-content is largely irrelevant to the Einstein debate, because Einstein undoubtedly captures and examines content, using a technique called deep-packet inspection.[23]

Also, I think it is worth noting that the distinction between content and non-content is increasingly losing its power as the dominant basis for setting privacy rules, including constitutionally derived rules, just as the distinction between personally identifiable information and "non-PII" is losing its salience as the basis for privacy regulation. The Supreme Court, well before the dawn of the digital age, adopted a binary view of the Fourth Amendment and in cases such as *United States v. Maryland* concluded that transactional data about communications was entitled to zero constitutional protection. Clearly, however, metadata is revealing and it is becoming ever more so, if only as a result of growing volume and analytic capability. Ultimately, I believe, the Court will have to reconsider its position that individuals voluntarily surrender all privacy interest in data that reveals the full pattern of their lives. It may be quite some time before that happens, but I believe that a policy according zero sensitivity to non-content is built on fragile ground.

The Solution Space

The answer, it seems, is to combine the best of both—the knowledge and agility of the private sector in understanding and responding to threats against its own systems, and the special insight that the government may have through its intelligence operations—making moot the debate over whether a government apparatus such as Einstein should be inserted into the private networks to monitor private com-

---

23.   The EINSTEIN 2 capability enables analysis of network flow information to identify potential malicious activity while conducting automatic full packet inspection of traffic entering or exiting U.S. Government networks for malicious activity using signature-based intrusion detection technology.

The Comprehensive National Cybersecurity Initiative (http://www.whitehouse.gov/cybersecurity/comprehensive-national-cybersecurity-initiative).

munications. And, in fact, that is what has been done, in what is called the Defense Industrial Base (DIB) Pilot.

For years, efforts to improve the security of privately owned and operated communications networks have been stymied by a conundrum: The government, particularly the National Security Agency (NSA), is presumed to have special knowledge of where cybersecurity attacks come from and what they look like. But the NSA has always maintained that its information-identifying cyber attacks could not be disclosed to the private sector without compromising sensitive intelligence sources and methods. This is what led to calls to insert government monitoring boxes deeper into private networks.

In the past year, there has been a major breakthrough on this gridlocked situation: The NSA and several major Internet service providers agreed that the private companies could, in fact, receive cyber-attack signatures from the NSA and use them to improve those companies' cybersecurity defenses while at the same time securing the information against compromise.[24]

The arrangement also overcame another conundrum: If the companies were required to report to NSA, or another government agency, on the communications that were identified and intercepted using government-supplied signatures, that would constitute a wiretap. In an all-too-rare display of pragmatic priority setting, the NSA said that it was willing to forgo any feedback from the companies on their use of NSA insights. (The Pilot permitted providers to report back, but did not require it.) The goal, all parties concluded, was to enhance the private sector's ability to defend its systems and its customers, not to perform backdoor wiretaps.

Having the Internet carriers monitor their networks using NSA data is an elegant solution to the long-standing problem of how to apply the government's special expertise while avoiding domestic surveillance by the government.

---

24. Ellen Nakashima, *NSA Allies with Internet Carriers to Thwart Cyber Attacks against Defense Firms*, WASH. POST (June 16, 2011), *available at* http://www.washingtonpost.com/national/major-internet-service-providers-cooperating-with-nsa-on-monitoring-traffic/2011/06/07/AG2dukXH_story.html.

However, while the NSA was developing this approach, the Administration was going in a different, far less desirable direction. Instead of helping private-network operators become better at the job they are already doing, the administration has sent legislation to Congress that would create a blanket exception to all privacy laws, allowing the network operators to share information about any and all communications with the government and placing responsibility in the government to do the analysis. The administration's proposal could result in a flood of private traffic flowing to the government.

Conclusion

The government should not be the central analysis point for Internet traffic, not only for the obvious civil liberties reasons but also from the standpoint of cybersecurity effectiveness. Even assuming that the government does have some specialized knowledge of attack methodologies that the private sector lacks, the government has neither the knowledge of those private networks nor the agility to act quickly enough to defend them when needed.

Einstein 3.0: Reply to Jim Dempsey

Paul Rosenzweig

Though styled as a "rebuttal," I prefer to think of this offering as a continuation of an ongoing discussion with my colleague Jim Dempsey. There is, after all, much we agree on and our disagreements are, as I think we both recognize, more in the matter of judgment than in any sense absolute.

To begin with, I take it from Jim Dempsey's general silence on the issue that he agrees with the overall structure of my legal analysis of Einstein 2.0, 3.0, and the prospective 4.0 under current standards. While I certainly agree that *Miller* and *Smith* are old law from the dawn of the computer era, and while I also agree that the content/non-content distinction has a tendency to break down at the margins, those legal constructs remain governing law today. It may be that the construct defines "antique privacy" (as I've called it elsewhere), but I see no reasonable prospect of change any time in the near future.

And so, I am left to apply existing law, and I come to the conclusion that any consensual deployment of a government program onto a private-sector network will be legal under current standards. What remains for discussion is whether it is wise and if it is done how it ought to be constructed. Here, Dempsey's reference to and reliance on the Defense Industrial Base (DIB) Pilot as a good model nicely illustrates the limits of his "private sector first" theme and our modest disagreements.

First, if it really were true that the federal government had nothing to add to cybersecurity, then the DIB Pilot would be unnecessary. That neither the private sector nor the government thinks it so is at least some evidence that the government (and more particularly, the NSA) is a "value added" contributor to cybersecurity in the private sector. Jim Dempsey and I can agree (and we do) that the private sector is more nimble than government and that in our capitalist society the private sector ought to be the cybersecurity provider of first choice. But my claim is a much more limited one—not that the federal government is always better, but only that, in some cases, it can be a superior provider of information or technology.[25]

Thus, with respect, I don't think this is really just an efficiency or efficacy debate; we both agree that only efficient and effective solutions should be chosen. Where we part ways (albeit gently) is in Dempsey's apparent belief that the private market will always be more effective. To my mind that is usually true, but it is not always so. After all, when Google was hacked by the Chinese in Operation Aurora, it turned to the NSA for assistance in analyzing the intru-

---

25. A recent internal DoD study suggested that the DIB Pilot's effectiveness has been mixed, with some success and some failures to meet expectations. Notably, in this test phase only two of 52 discovered incidents relied on classified NSA signatures. *See* Ellen Nakashima, *Cyber Defense Effort Is Mixed, Study Finds*, WASH. POST, Jan. 12, 2012, *available at* http://www.washingtonpost.com/world/national-security/cyber-defense-effort-is-mixed-study-finds/2012/01/11/gIQAAu0YtP_story.html.

sion.[26] At least to some degree, cyber market leaders such as Google recognize the value of federal cyber assistance.

More fundamentally, however, the DIB Pilot shows how a fear of government intervention can have a tendency to hamstring the effectiveness of our collective approach to cybersecurity. As Dempsey notes, to overcome lingering privacy-protective suspicions of its motives, NSA agreed to limited, voluntary feedback from the DIB Pilot partners on the effectiveness of the NSA-provided threat signatures.[27] This was a pragmatic decision—as the enhanced security provided to the DIB was sufficiently important to NSA that it was willing to avoid the political controversy that might arise if it was suspected of "perform[ing] backdoor wiretaps," as Dempsey puts it.

But let us make no mistake: This form of self-limitation is not intended to enhance effectiveness or efficiency. To the contrary, there can be little doubt that NSA's response and its ability to provide signature threat information to all the members of the DIB Pilot is enhanced to the extent it is able to incorporate information on effectiveness derived from the members' experience. Nobody who wants to foster product improvement purposefully constructs an OODA

---

26. Ellen Nakashima, *Google to Enlist NSA to Help It Ward Off Cyberattacks*, WASH. POST, Feb. 3, 2010, *available at* http://www.washingtonpost.com/wp-dyn/content/article/2010/02/03/AR2010020304057.html?hpid=topnews.

27. The precise extent and nature of this voluntary feedback are unclear from official records. DoD says only that companies are "asked to report network incidents." According to the recent Privacy Impact Statement published by DHS,

> The [ISP] may, with the permission of the participating DIB company, also provide some limited information about the incident to US-CERT sufficient to capture the fact of occurrence. US-CERT may share the fact of occurrence information with DoD pursuant to existing US-CERT procedures in an effort to increase DoD's understanding of the threats to their critical assets that reside within the DIB companies' networks and system. The [ISPs] may voluntarily choose to send US-CERT information related to cyber threat indicators or other possible known or suspected cyber threats.

PIA for Joint Cybersecurity Services Pilot, Jan. 13, 2012. *See* http://www.dhs.gov/xlibrary/assets/privacy/privacy_nppd_jcsp_pia.pdf.

loop[28] without a feedback component. Where we could, in my judgment, lawfully have expanded NSA's role through a "strong consent" model, we have, instead, deliberately chosen to provide only a limited feedback loop for political reasons.

This course will inevitably make us less effective. As a pragmatic judgment of the politically possible in America today, it is probably a wise decision. As a policy for effective cybersecurity, it is the second-best option.

Einstein 3.0: Reply to Paul Rosenzweig

James X. Dempsey

In his rebuttal, my colleague Paul Rosenzweig has moved very far indeed from his initial effort to defend the proposition that the federal government should monitor private sector communications through insertion of an Einstein 3.0 or 4.0 into the networks owned and operated by the private sector. Now his main point, which I accept, is that it would be useful to have a feedback loop when the private sector uncovers cyber attacks using information provided by the government. If there were another round of comments, we could probably agree on how to narrowly scope that data flow.

This "debate" has come a long way from the vision propounded by some (and never fully advanced by Rosenzweig himself) that the initial analysis of private-to-private communications should be performed by black boxes owned and programmed by the government and that the secondary analysis should be performed by a government-run center. In retrospect, the concept seems fantastical; the government has a hard enough time defending its own networks. Even without the serious civil liberties concerns, there seems to be growing recognition that the government is ill-equipped to fulfill a centralizing role.

---

28. OODA stands for "Observe, Orient, Decide and Act." An OODA loop is a conceptual means of examining commercial operations or learning processes (originally designed by the military), and always features a feedback component, so one can learn by doing.

Thus, we seem to have agreement that cybersecurity monitoring of private-to-private communications, including the initial analysis of hits and anomalies, should be performed by the private-sector owners and operators of critical infrastructure. They are in the best position to understand their own networks, to identify attacks, and to take preventative and remedial action. The major communications service providers, financial institutions, and others already have years of experience in monitoring their systems for security purposes. They have been steadily improving their defenses. The government may have some special insight into threats and vulnerabilities, based on its foreign intelligence and information security activities, but policy reforms should aim to facilitate sharing that insight with the private sector in a secure fashion, not to have the government usurp the private sector's role.

Rosenzweig is concerned that distrust of the government may impede the effectiveness of a cybersecurity information-sharing effort by curtailing the flow of data back to the government. However, rather than accepting distrust as an impediment to effectiveness, we should seek to develop and leverage trust as an enabler of effectiveness. To develop that trust, we must ensure that the government is not using private-sector monitoring as a de facto wiretap. Instead, let's carefully define the categories of data that can flow to the government, such as attribution information on suspected attacks and data about apparent probes in advance of attacks.

While Rosenzweig and I have come very close to agreement on the bottom line—empower, don't usurp the private sector in its role defending privately owned and operated networks—I cannot let this exchange close without a further comment on the question of consent. Rosenzweig hinged much of his initial legal analysis on consent, arguing that the government could intrude very deeply into private communications streams if service providers could gain the consent of their customers.

I think it is best to warn policy makers and private stakeholders to be cautious of consent. A narrow reading of current law would indicate that consent excuses all intrusions, but a countertrend is under way, and those building cybersecurity policies on a broad read-ing of consent may find their foundation shaky. There is a growing recognition that the notice models of the past are not working and that consent too easily obtained is not adequate to override privacy expectations. For several years, senior officials at the Commerce Department and the Federal Trade Commission have been warning that the notice-and-consent model for data collection is not fair to consumers. In recent enforcement actions, the Commission has gone after companies that collected consumer information in unexpected ways even if they obtained consumer consent. In related Fourth Amendment contexts, some courts have questioned the view of con-sent underlying the *Miller* and *Smith* cases. Carried to an extreme, consent would eliminate all privacy. For communications to the gov-ernment itself, and with defense contractors, consent is reasonable as a basis for monitoring (although even there oversight is needed). For the broader Internet, it is far better to base cybersecurity improve-ments on better design and accountable information sharing.

Chapter Seven
CALEA: What's Next?

Tony Rutkowski

Susan Landau

Introduction

T he Communications Assistance for Law Enforcement Act (CALEA) was adopted in October of 1994 to preserve the ability of law enforcement officials to conduct electronic surveillance—a capability that had been put at risk as industry rolled out new digital technologies and wireless services. Enacted over the objection of privacy and technology advocates who feared that it would constrain the development of new technology, CALEA requires telecommunications carriers to modify their equipment, facilities, and services, wherever reasonably achievable, to ensure that they are able to comply with authorized electronic surveillance.[1]

What followed was nearly eight years of dialogue and litigation between industry and law enforcement agencies to determine how Congress's broad mandate would be applied to a host of specific technologies. In June of 2002, the Federal Communications Commission (FCC) approved a detailed set of surveillance standards for telecommunication carriers.

---

1. Note that CALEA applies to communications content rather than monitoring the general activity of specific IP addresses, for which the government requires the equivalent of a pen register or trap-and-trace device (rather than a warrant). For more, *see* Patriot Debates (2004) *available at* (http://www.steptoe.com/publications/373b.pdf).

But technology did not stand still. The widespread deployment of broadband Internet-based "packet mode" services allowed Voice-over-Internet-Protocol[2] (VoIP). VoIP opened up the possibility that individuals could talk online to others—around the world or down the block—without ever making a traditional phone call. This prospect led law enforcement agencies to ask the FCC to clarify who is a "telecommunication carrier" and therefore subject to the requirements of CALEA. The agencies also requested clarification of the specific CALEA obligations of broadband access and phone services.[3] Industry and privacy groups resisted.

In August of 2005, the FCC ruled that CALEA applied to private (facilities-based) broadband Internet access providers.[4] It also took on the VoIP problem, drawing a line between VoIP providers who tied their customers' communications into the standard phone network (the Public Switched Telephone Network or PSTN) and those who did not. Any service that did not allow calls to and from the PTSN, the FCC held, was *not* subject to the provisions of CALEA, because "non-switched" service did not qualify as a "substantial replacement" for ordinary phone service. In practice, this meant that CALEA applied if the VoIP service allowed the customer to both make and receive calls to or from people with ordinary phones. The FCC required compliance by May of 2007.

But no sooner was the distinction between switched and non-switched services adopted than technology shifted again. Dependence on the PSTN began to decline. If the service was cheap enough,

---

2. VoIP is an acronym for "Voice over Internet Protocol" and refers to voice or multimedia communications transmitted over the Internet.

3. There was a difference in opinion between the FBI and industry representatives as to which specific surveillance capabilities CALEA facilitated. *See* Patricia M. Figliola, *Digital Surveillance: The Communications Assistance for Law Enforcement Act*, Cong. Reporting Service (2007), *available at* http://www.fas.org/sgp/crs/intel/RL30677.pdf (Aug. 17, 2011).

4. *See* First Report and Order and Further Notice of Proposed Rulemaking (First RO), in ET Docket No. 04-295, FCC 05-153, Sept. 23, 2005, at 13. 4 Valerie Caproni, FBI General Counsel, Statement before the House Judiciary Committee Subcommittee on Crime, Terrorism and Homeland Security, Feb. 17, 2011, at 1.

many customers did not care whether it handled calls to and from regular phones. As this attitude spread, neither the FCC nor the FBI could be confident that CALEA applied to every substantial VoIP service. Indeed, today, many online communications services carry plenty of live communications without offering full connection to the PSTN—or the wiretap capabilities that law enforcement wants.

The rise of such services has led the FBI to argue that its intelligence capabilities are gradually "going dark," and that it needs new authority to address these CALEA-exempt areas. In the words of its General Counsel,

It is no longer the case that the technology involved in communications services is largely standard. Now, communications occur through a wide variety of means, including cable, wireline, and wireless broadband, peer-to-peer and VoIP services, and third party applications and providers—all of which have their own technology challenges. Today's providers offer more sophisticated communications services than ever before, and an increasing number of the most popular communications modalities are not covered by CALEA.

> . . . . [S]ome providers are currently obligated by law to have technical solutions in place prior to receiving a court order to intercept electronic communications but do not maintain those solutions in a manner consistent with their legal mandate. Other providers have no such existing mandate and simply develop capabilities upon receipt of a court order. In our experience, some providers actively work with the government to develop intercept solutions while others do not have the technical expertise or resources to do so. As a result, on a regular basis, the government is unable to obtain communications and related data, even when authorized by a court to do so. We call this capabilities gap the "Going Dark" problem.[5]

---

5. *Id.*

For critics, though, history tells a different story. Some question the entire effort to make law enforcement interests a constraint on technological development. Others argue that pursuing an ever-broader scope for CALEA will impose crippling financial and technological burdens on innovative companies. Finally, some fear the cybersecurity implications of building wiretap "doors" into communication devices, asserting that the benefits of easier wiretapping do not outweigh the danger of third parties using the same door illegally.

In a Cloud-Based Mobile World: Providing Needed Forensics

Tony Rutkowski

CALEA is a term with two entwined meanings and often wrapped with a liberal dose of Washington politics and anti-government paranoia. The term simply refers to requirements that have long existed in essentially every jurisdiction in the world that network operators and service providers be capable of handing over certain types of forensic criminal evidence in a timely fashion to lawful authority when presented with a production order. CALEA also refers to the Communications Assistance for Law Enforcement Act of 1994 and its subsequent implementation pursuant to regulations of the Federal Communications Commission (FCC). Although the answers to the question "What's next?" differ in the short run depending on which CALEA meaning is addressed, in the long term the directions are likely the same.

The world of global information communication networks continues to expand and evolve significantly. The need for forensic information for public safety, including analysis and evidence—both criminal and civil—has expanded even more. If anything, the past decade of ubiquitous, frequently mobile, always-connected and always-on user devices and behavior, combined with exponentially more complex software defined devices and services, and with government deregulation, have resulted in exponentially scaling global vulnerability exploitation and crime subject to legal remediation. The rapid growth of "cloud" virtualization services exacerbates the challenges.

Chapter Seven: CALEA: What's Next?

CALEA—like its sister law enforcement assistance provisions in countless jurisdictions worldwide, including treaty instruments such as the Convention on Cybercrime, the European Union Data Retention Directive, or U.S. eDiscovery Rule 26—are all designed to be generic, constantly evolving means to provide essential forensics. Increasingly today, the same means are necessary to protect networks and services from cyber attack. All these needs have produced a robust global supporting industry, technical standards bodies, and new government mandates for needed capabilities.

The forensic handover capabilities are used especially for the world's most ubiquitous, most used, and rapidly expanding communications infrastructure—mobile cellular networks—now numbering more end-user terminals than humans on the planet. That infrastructure is becoming extensively "gatewayed" with the somewhat smaller network serving about a third as many people—the popular Internet. They both ride together with many specialized networks on top of a lower network substrate of ever less expensive bandwidth, processing, and storage, all increasingly provided through virtualized cloud services resident at massive data centers distributed worldwide. All of this communications infrastructure and the services provided are subject to lawful interception (LI) and retained data (RD) requirements throughout the world. Within the United States, CALEA facilitates the availability of LI capabilities.

The LI capability requirements necessarily evolved significantly to keep pace with changing infrastructure and services. To accomplish this, judicial and law enforcement authorities and intelligence agencies interact constantly around the world with network operators, service providers, and equipment vendors to examine emerging challenges, determine requirements, develop "handover" standards, and implement capabilities.

The "What's next?" paradigm has long been under way. However, especially in the United States, there remain major gaps in both the law and implementations that U.S. government agencies refer to as the "Going Dark" problem.

[S]ome providers are currently obligated by law to have technical solutions in place prior to receiving a court order to intercept electronic communications but do not maintain those solutions in a manner consistent with their legal mandate. Other providers have no such existing mandate and simply develop capabilities upon receipt of a court order. In our experience, some providers actively work with the government to develop intercept solutions while others do not have the technical expertise or resources to do so. As a result, on a regular basis, the government is unable to obtain communications and related data, even when authorized by a court to do so.[6]

Challenges

Forensics challenges faced worldwide by government authorities are fairly similar. Everyone is dealing with rapid changes in technology, network and services operations, and user practices and devices. What seems not understood by many not familiar with CALEA is that there are no "backdoors" required. There are not even specific technical mechanisms required. The technical standards that exist are written in the form of a single "interface" that must be supported by the service provider for authorities, at which requests are received and evidence furnished in a uniform structured format. In simple terms, the government authority is provided a permanent address at which it can present a court order, and request specific evidence associated with a customer, and the service provider then "hands over" that evidence in a timely fashion. It is, in fact, a "front door." Indeed, the interface is generally referred to as a Handover Interface for obtaining either real-time or delayed forensics, depending on the nature of the evidence sought and the order. The forensics can include either (1) communications-related network signaling, such as when and to whom some communication occurred, or (2) the actual communication content.

---

6. *Id.*

CALEA—as the U.S. Court of Appeals underscored in its 2006 decision on appeal from the FCC's *CALEA Order* on Internet Protocol forensics—was written to be technology-neutral and evolve. The forensic requirements are generic ones that can be relatively flexibly applied as new technologies and services evolve. However, there is one significant legal sticking point raised in the FBI's "Going Dark" plea. Essentially every nation in the world has a CALEA-like requirement. Incredibly, the United States is alone among the nations of the world in applying a half century of FCC categorization known as "information services" (which the FCC itself doesn't even use anymore) as a limitation on CALEA's applicability.

The exception occurred because a powerful information-provider lobbying community (which also doesn't exist anymore) managed in 1994 to convince Congress to insert the ancient categorization as an exception into the CALEA statute. This has resulted in the ensuing years in the inanity of some providers attempting to escape CALEA requirements by characterizing themselves as "information services," as well as the FCC and the judiciary engaging in Talmudic exercises to deal with an absurd statutory provision harming the national public safety as a result from Washington lobbying excesses.

CALEA technical and operational challenges are difficult but largely have solutions, given sufficient resources. The following list summarizes some of the principal challenges.

a. Maintain handover capabilities for legacy systems for LI and Retention of Data (RD), while evolving these capabilities to address the ever increasing complexity and technology advancements within the global communication services market.

b. Provide for the ability to deal with scaling transport bandwidths and dynamic terminal access.

c. In a world of application-based networks, provide for a means for implementing LI and RD capabilities.

d. Provide for handover capabilities for Cloud Computing facilities and service providers, including "social network services."

e. Provide for the implementation of handover capabilities for machine-to-machine (M2M) and peer-to-peer (P2P) communications.

f. Provide means to deal with the increasing encrypting of content.

g. Enhance trust in network infrastructures and services, including those supporting LI/RD requirements.

Because these are global needs, a global market exists for the solution capabilities and a significant number of industry players are involved. In many instances, these capabilities are also needed for protecting networks and users, which allows for significant product platform synergies. Indeed, when the crime itself is a cyber attack, the associated forensics from those attacks are made available to law enforcement authorities, as well as cybersecurity remediation centers.

So, What's Next?

CALEA requirements and capabilities in large measure exist and are being driven by industry and government users worldwide. However, like much of the activity and innovation in the communications networking sector today, the growth, resources, and innovations today are increasingly offshore. The LI/RD industry collaborative activity and trade shows outside the United States are larger, more numerous, and exhibit more innovations than the one remaining event in the United States. The new and innovative methods and techniques evolving out of that scaling global work will benefit the United States in meeting its technical and operational challenges.

The non-technical side of the U.S. "Going Dark" problem is another matter. In a rational political process, the United States, like most other nations, would require the capabilities and instantiate a funding mechanism for the relevant government agencies to put necessary LI/RD capabilities in place. However, CALEA is in the same political realm as the debt ceiling, and the U.S. political process is not well suited to acting on some matters even when they are essential to the national interest. It has been four years now since the

Department of Justice requested trivial CALEA action at the FCC. No known enforcement of existing provisions seems to have occurred. Action by the Commission remains problematic.

Up on the Hill in Washington, one would hope that, just maybe, CALEA's inane half-century-old "information service exception" albatross that is unique to the United States could at last be removed from the neck of the law enforcement community by Congress. If it wanted to gain extra points, Congress could also demonstrate some positive response to law enforcement's requests for the additional tools needed to deal with today's challenges.

Legally Authorized Wiretapping—but Not Communications Insecurity

Susan Landau

In placing the Federal Bureau of Investigation (FBI) in the role of determining telecommunications surveillance standards, CALEA marked a radical departure for electronic surveillance law. The philosophy behind CALEA was that service providers should build wiretapping capabilities into communications systems. But that solution creates risk, a serious issue, given the increasing number of cyber attacks and cyber exploitations being conducted against the United States. To understand the problem that embedding surveillance tools into communications infrastructure creates, I will start with a brief discussion of the technology of communications.

Despite FBI claims in the early 1990s that wiretapping was becoming difficult, the fact is that wiretapping communications on the Public Switched Telephone Network, or PSTN, is a straightforward effort. That is because the telephone network is a circuit-based system: Each call occupies a fixed "circuit" for the duration of the call. Parts of this circuit are predictable. A call goes from caller to callee by first accessing the phone central office nearest the caller. It is trivial to determine at which central office the tap should be placed; it is activated when a call is made. Wiretapping cellular calls is somewhat more complicated, as mobility affects the caller's location. Incoming cellular calls are routed to a Mobile Switching Center. That is where a tap can be placed. Tapping outgoing calls when the caller is roaming is more complicated.

Such techniques generally break down in the Internet, which is a packet-switched network. In the Internet, communications are broken into small "packets," each of which may take its own path from sender to recipient. It is difficult both to predict where the sender will be and which route packets will take. Thus, there is no clear way to determine where a wiretap should be placed for an Internet-based communication.

The Federal Communications Commission (FCC) "solved" the VoIP problem by applying CALEA only to those communication cases in which the architectures of the PSTN and the Internet were similar. The FCC ruled—and an Appeals Court ruling confirmed[7]— that the law applied to broadband Internet access providers and providers of interconnected VoIP services, VoIP capable of sending and receiving calls from the PSTN. Because these instances connect the caller from fixed, predetermined locations, installing wiretaps is much like installing a tap on a PSTN communication.

In recent years, law enforcement has faced an increasingly complex surveillance situation. Investigators encounter a multitude of communication services and providers (including those of social networking sites) and a plethora of architectures and operating systems. The result: Legal authorization for a wiretap may not result in the ability to actually conduct the surveillance.

At the same time, other communications changes have simplified law enforcement efforts. New real-time communications systems have created a vast array of services and data available to law enforcement and national security investigators. In many instances, cases can be solved faster and with fewer resources than previously. For example, using location information, the average time for the U.S. Marshals Service to locate fugitives has dropped from forty-

---

7. American Council of Education v. Fed. Commic'ns Comm'n, 371 U.S. App. D.C. 307; 451 F.3d 226 (2006); U.S. App. Judge Edwards dissented, pointing out CALEA's exemption for "information services."

two days to two.[8] Through pen registers and trap-and-trace orders, investigators can map out the organization of a conspiracy simply based on the web of communications connections. This information can be obtained at a fraction of the time and cost of surveillance teams.

Yet just as in the early 1990s, law enforcement says it is "going dark" because of communications changes. The FBI has cited encryption and peer-to-peer communication systems[9] as particular difficulties. The FBI seeks to expand CALEA to network communications, although the exact type of extension being proposed remains unclear as of this writing.[10]

Because the PSTN and the Internet are vastly different networks, solutions are not that simple. The PSTN and the Internet often deliver the same type of content, although the Internet serves as a far more complex and diverse communications network than does the PSTN. Technologists describe the architectural difference as the PSTN having dumb terminals—phones—and smart networks designed for transmitting high-quality voice, while the Internet is designed for smart endpoints—computers—and as a less smart network. In fact, there is much clever engineering in the switches and routers that direct packet flow on the Internet. The matter that directly concerns us is these endpoints, which on the Internet have vast computational power. The latter enable the nimble applications that are the driver of Internet innovation. But that very nimbleness—the endhost computer programmability—is the root of many of the Internet's vulnerabilities.

Lack of security means that the network provides an inexpensive and extremely useful way for both organized crime and other nation-

---

8. Susan Landau, SURVEILLANCE OR SECURITY? THE RISKS POSED BY NEW WIRETAPPING TECHNOLOGIES, MIT PRESS (2011) at 100.

9. A peer-to-peer network is one in which no particular machines take charge of communications, and nodes can be the initiator or recipient of a message, or simply pass the message through to a different machine (Landau, at 18).

10. Statement of Valerie Caproni, General Counsel, FBI, Before the Subcommittee on Crime, Terrorism, and Homeland Security, U.S. House of Representatives (Feb. 17, 2011).

states to steal money—and more. This is an increasingly serious national security problem. According to Deputy Secretary of Defense William Lynn III, the theft of U.S. intellectual property—research data and software, confidential business plans and classified information—may be the "most significant cyber threat that the United States will face over the long term."[11] Thus, the proposed FBI direction is highly problematic. After all, CALEA mandates building architected security breaches into communications infrastructure. This puts society at risk. When a communications switch or application is vulnerable, all communications using it are in danger.

Such risks are not theoretical.[12] In 2004 and 2005, parties unknown accessed a Vodafone Greece switch equipped with wiretapping capabilities that had been designed to accommodate law enforcement requirements. The result was that 100 senior Greek officials, including the prime minister, were wiretapped for 10 months.[13] Between 1996 and 2006, more than 6,000 Italians, including judges, politicians, celebrities, and sports figures, were similarly wiretapped, presumably for bribery and blackmail purposes (the case remains in court).[14] In 2010, an IBM researcher found that a Cisco switch designed to accommodate law enforcement wiretapping could be spoofed, allowing unauthorized parties to initiate eavesdropping.[15] It is true that law enforcement faces complex new communications technologies, but a broad-brush approach that introduces vulnerabilities into these is not the way to solve law enforcement's problem.

One should start by a careful analysis of the different communication architectures. These fall into three broad types of systems:

---

11. William J. Lynn III, *Defending a New Domain*, FOREIGN AFFAIRS (September/October 2010).

12. The text that follows appeared originally in Susan Landau, *Getting Wiretapping Right*, HUFFINGTON POST (April 5, 2011), *available at* http://www.huffingtonpost.com/susan-landau/getting-wiretapping-right_b_844733.html.

13. Vassilis Prevelakis & Diomidis Spinellis, *The Athens Affair*, IEEE SPECTRUM (July 2007), 18–25.

14. Piero Colaprico, *'Da Telecom Dossier sui Ds' Mancini parla dei politici*, LA REPUBBLICA (January 2007).

15. Tom Cross, *Exploiting Lawful Intercept to Wiretap the Internet*, BLACK HAT DC (February 2010).

- Centralized systems such as the PSTN;
- Internet communications systems that rely on central data repositories (e.g., Gmail[16]); and
- Purely peer-to-peer communication systems, such as Skype.[17]

This categorization clarifies where law enforcement solutions may lie. The first set of communications is already covered by current surveillance technologies. Although the second category is a new form of communications, it does not present technical problems for wiretapping. Any centralized communication system necessarily has an access point for surveillance. All that is required is for law enforcement to set the groundwork for surveillance prior to serving an actual warrant. That means law enforcement has to determine the technological way to accomplish the surveillance and to ensure that systems are in place to be able to do so once a warrant is served.

Peer-to-peer communications lacking a centralized data repository do present surveillance difficulties. But even here there are solutions. The FBI has shown its ability to download malware onto target systems to enable keylogging and other forms of surveillance. Although such approaches may need to be individually developed, and may be more complex and expensive, they can be pursued in cases that are otherwise difficult to handle.

Currently, law enforcement approaches wiretapping on a case-by-case basis, with the result that new communications technologies can sometimes thwart timely taps. Instead, the FBI needs to proactively conduct research into "There's a new communications technology; what is the method for conducting surveillance (once there is a warrant)?" Such techniques should be on the shelf for use when needed. Otherwise, time is lost during investigations developing surveillance techniques (and contacts at the company). That only benefits the bad

---

16. Sergey Brin & Lawrence Page, *The Anatomy of a Large-Scale Hypertextual Web Search Engine*, at http://infolab.stanford.edu/~backrub/google.html.

17. Skype, *What Are P2P Communications?*, at https://support.skype.com/en-us/faq/FA10983/What-are-P2P-communications [last viewed Aug. 29, 2011].

guys. I favor increased funding for developing an active research component for the FBI's "Going Dark" program.

Another difficulty that state and local investigators face is trouble determining how to wiretap new communications technologies; they lack resources to handle each new system.[18] That's where the FBI can step in. Currently, such information sharing is ad hoc, and that again only benefits the criminals. The most recent statistics from the Administrative Office of the U.S. Courts show that state and local police conduct 72 percent of law enforcement wiretaps.[19] Thus, a well-designed information-sharing system for wiretapping techniques would have immediate beneficial impact.

That's where I would go next. Wiretapping is an important investigative tool, but solving law enforcement's difficulties should not come at the expense of increasing cybersecurity vulnerabilities. Building further surveillance capability into our communications infrastructure creates long-term risk—and that is exactly the opposite of what we should be doing.

CALEA: "What's Next?"—A Rejoinder

Tony Rutkowski

It would appear that Susan Landau and I strongly agree on the needs and objectives of "What's next?" for CALEA. Congress should provide law enforcement with the needed resources to implement the capabilities. Those capabilities should not exacerbate the already massive vulnerabilities of the nation's network and services infrastructure.

Our most obvious differences concern the legal and technical facts involving CALEA's implementation and the larger world of contemporary infrastructure and lawful interception tradecraft. Both

---

18. Statement of Mark Marshall, President, Int'l Chiefs of Police, Before the Subcommittee on Crime, Terrorism, and Homeland Security, U.S. House of Representatives (Feb. 17, 2011).

19. Administrative Office of the U.S. Courts, Dep't of Justice, 2010 Wiretap Report, tbl. 4.

of these areas are generally not well understood outside of the eso-teric, highly specialized, and generally compartmentalized commu-nities involved. A few examples are provided.

It is asserted that CALEA places the "FBI in the role of deter-mining telecommunications surveillance standards." In fact, it does not. The Bureau only describes generic requirements that are then implemented by service providers subject to CALEA. It turns out that those generic requirements are similar to those that exist world-wide, and are indeed coordinated globally, because network infra-structures and services today are global. The standards and solutions that industry develops and implements for operators and providers are global as well. Individual service providers can also "roll their own" solutions, frequently through third-party CALEA service pro-viders, and many of them do. So the FBI's role in practice tends to be rather minimal. It is also similar to most of the other law enforce-ment and public safety agencies worldwide that are all facing an onslaught of cyber crime.

It is asserted that today's network infrastructure consists of three disparate worlds consisting of "centralized systems such as the PSTN, Internet communications systems that rely on central data reposito-ries, and purely peer-to-peer communication systems." The reality is that the services and infrastructures largely exist on a common net-work substrate that decades ago became packet-based, and they all possess data repositories. It is all a mesh of transport capabilities, computer processing, and storage that are increasingly becoming aggregated as cloud data centers. As this has all occurred, the com-munities that deal with both network forensics and infrastructure protection have developed the technical capabilities to meet their needs. The problem is that whether it is cybersecurity protection or law enforcement evidentiary needs, what is required does not get ubiquitously implemented without mandates and money.

It is asserted that "CALEA mandates building architected secu-rity breaches into communications infrastructure." In fact it does not. As explained above, CALEA mandates an available "front door" interface at which the 15,000 different law enforcement agencies in the country can deliver court orders for forensic requests in some reasonably consistent way and get those forensics provided in a manner they can use. How the providers technically implement the capabilities in their networks is their choice. In practice, this is done today via overlay acquisition mechanisms that are maintained separately from the network and introduce no vulnerabilities to the network itself. Increasingly, the same capabilities are used to protect the network from threats to itself and provide needed forensics when attacks arise. Such overlay mechanisms also avoid imposing any constraints on innovating and evolving networks and services—another urban legend bandied about in conjunction with CALEA politics.

Proffered examples of researchers finding widespread vulnerabilities in network devices tend to be bogus when subjected to scrutiny. The well-traveled story of an "IBM researcher" occurred at a hackers conference, and the vulnerability was a well-known generic one associated with a network management protocol vulnerability that had been publicly reported years earlier. The protocol was not actually a vulnerability as implemented by vendors, and it was not mandated by CALEA. Ironically, this vulnerability is only one of tens of thousands of other network vulnerabilities reported that have nothing even remotely to do with forensics acquisitions. It makes for great tabloid reading and political bantering at congressional hearings, but the facts differ substantially from reality. As noted above, the forensics acquisition mechanism involved has also transitioned to other approaches today.

One of the unfortunate dimensions of congressional and FCC inaction in the lawful interception and data-retention fields not mentioned in the Going Dark dialogue is that the technology and expertise have grown significantly abroad while diminishing domestically. Europe and the Middle East collectively, for example, have many more forensics vendors today than the United States. Because the technology and expertise have applicability for cybersecurity, it may become a detriment to national security. Unfortunately, this also is an important component of "What's next?"

CALEA: "What's Next?"—Rejoinder

Susan Landau

I'm not surprised that Tony Rutkowski focuses on new capabilities for law enforcement instead of paying attention to old problems. If I were law enforcement, I'd try to ignore the security issues created by CALEA-type systems. I'd do the same if I were selling equipment to accomplish CALEA-type goals. But this attempted sleight of hand doesn't make security issues disappear. Consider Rutkowski's proposals in light of the risks.

Tools for automatic handover of data are of great concern. Rutkowski says these law enforcement interfaces are not a backdoor but a "front door." The problem, however, is that there is a door in the first place. Perhaps law enforcement and providers of surveillance equipment have not heard of Murphy's Law, insiders gone bad, and subvertible software, but you can be sure that our opponents—organized crime and other nation-states—have. They employ them regularly.

Rutkowski includes a proposal for evolving handover capabilities of data retention information—the transactional information of participants, time, and location of a communication. One wonders if anyone has been reading the newspapers of late.[20] Over a period of years, UK journalists illegally accessed the voicemail of princes, police, celebrities, sports figures, and ordinary people. The activities of the *News of the World* and related papers showed compromises of service providers and law enforcement. According to a member of Parliament, "There is clear evidence that in some cases rogue staff members [of mobile phone companies] sold information to investigators and reporters."[21] Reporters paid police officers to find people's locations through requests to service providers. Targets were located

---

20. Some text that follows originally appeared in Susan Landau, *Data Retention? News of the World Demos the Risks*, HUFFINGTON POST (July 21, 2011).

21. LES HINTON, U.K. PARLIAMENTARY BUSINESS (March 10, 2011), *available at* http://www.publications.parliament.uk/pa/cm201011/cmhansrd/cm110310/debtext/110310-0004.htm.

through the "pings" of their cell phones to adjacent towers.[22] Such techniques were only for "high-profile criminal cases and terrorism investigations,"[23] but payoffs and corrupt police subverted these rules. Motivation for acquiring communications—content or transactional information—is high.

That is also true in the United States. In 2007, the Department of Justice Office of Inspector General found numerous violations in FBI use of exigent National Security Letters[24] (NSLs requesting immediate access to phone records with statements that subpoenas would follow): no paper trail, but only a verbal request;[25] searches with dates and other specifics missing in the request (resulting in more information being supplied than ought to have been),[26] fishing expeditions to discover the circle of contacts of a potential target.[27] This was not the first time there has been trouble with unauthorized FBI use of electronic surveillance,[28] nor is the FBI the only investigative agency to have been found misusing this highly invasive tool.[29]

Rutkowski proposes extending CALEA's interfaces to transactional information. But when surveillance mechanisms are easy to turn on, the chance for misuse is high. The Athens break-in and FBI misuse of exigent letters demonstrate these risks.

Rutkowski says that CALEA handover requirements apply to peer-to-peer systems. That means building surveillance capability

---

22. John Burns & Jo Becker, *Murdoch Tabloids' Targets Included Downing Street and the Crown*, N.Y. TIMES (July 11, 2011).

23. Burns, *supra* note 3.

24. U.S. DEP'T OF JUSTICE, OFFICE OF THE INSPECTOR GENERAL, OVERSIGHT AND REVIEW DIV., A REVIEW OF THE FEDERAL BUREAU OF INVESTIGATION'S USE OF EXIGENT LETTERS AND OTHER INFORMAL REQUESTS FOR TELEPHONE RECORDS (January 2010).

25. Department of Justice, *supra* note 6, at 137–38.

26. *Id.* at 70.

27. *Id.* at 74.

28. *See, e.g.*, U.S. CONGRESS, SENATE SELECT COMMITTEE TO STUDY GOVERNMENTAL OPERATIONS WITH RESPECT TO INTELLIGENCE ACTIVITIES, FINAL REPORT OF THE SELECT COMMITTEE TO STUDY GOVERNMENTAL OPERATIONS WITH RESPECT TO INTELLIGENCE ACTIVITIES: SUPPLEMENTARY DETAILED STAFF REPORTS ON INTELLIGENCE ACTIVITIES AND THE RIGHTS OF AMERICANS: BOOK III, REPORT 94-755 (April 23, 1976).

29. *See, e.g.*, U.S. Congress, *supra* note 9, and James Risen & Eric Lichtblau, *Officials Say U.S. Wiretaps Exceeded Law*, N.Y. TIMES (April 15, 2009).

into the endpoints or communications applications—another way to build insecurity into systems.

Rutkowski suggests there be "means to deal with the increasing encryption of content," but he goes against the government's decision in 2000 to enable wide deployment of cryptography.[30] That decision was made because even if some interceptions are unintelligible, widespread use of encryption increases U.S. security. That remains true.

Rutkowski argues that the rest of the world is requiring surveillance capabilities and building technologies, and that the United States will be left behind if it doesn't. He misses critical points. Much of the world doesn't have a Fourth Amendment or care about providing the communications privacy to citizens and political activists that our laws require. The argument to apply CALEA to insecure communications networks invokes short-term security—arrest this criminal—and is oblivious to long-term risks. Proposals to expand CALEA ignore the impact from the United States being more highly networked than much of the world. We face much more risk from surveillance capabilities being built into communications switches or applications than our opponents do. What better way to undermine U.S. security than through simplifying other nations' abilities to spy on domestic communications? Expanding CALEA's scope is, in fact, a dangerous proposition.

The FBI floated the idea of CALEA expansion in September of 2010. Notably, neither the Department of Defense nor the Department of State has come forward to support this initiative. It seems that they understand the risks involved in the FBI proposal better than the Bureau does. It's time to do the simple, easy things that don't threaten security, such as improving the FBI's quality of information sharing on cell phone forensics with state and local law en-

---

30. *See* KENNETH DAM & HERBERT LIN, CRYPTOGRAPHY'S ROLE IN SECURING THE INFORMATION SOCIETY, National Academies Press (1996) and Dep't of Commerce, Bureau of Export Admin., 15 C.F.R. pts. 734, 740, 742, 770, 772, and 774, Docket No. RIN: 0694-AC11, Revisions to Encryption Items. Effective Jan. 14, 2000.

forcement and developing a high-quality research capability on surveillance in new communications technologies. But let's leave aside the risky CALEA expansion. That will only create serious long-term dangers for the United States.

CALEA: "What's Next?"—Reply to Professor Landau's Rejoinder

Tony Rutkowski

Professor Landau's rejoinder creates a "Bizarro World" pastiche consisting of snippets from the world press about unlawful access to communications combined with other odd constructs that have little or nothing to do with the subject matter of CALEA provisions in the United States. For example, she argues that failure of the DOD and the State Department to publicly support DOJ CALEA arguments during congressional hearings means DOJ's proposals are "risky." Huh, really?

The challenge here in developing a reply to the Landau rejoinder is the utter lack of any apparent cognizance of what CALEA is. This is made plain in fictional phrases such as "CALEA-type systems" or "subvertible software" or "transactional information." Perhaps the *pièce de résistance* is "Rutkowski says these law enforcement interfaces are not a backdoor, but a 'front door.' The problem, however, is that there is a door in the first place." Inquiring minds might ask just how law enforcement is supposed to request the handover of evidence with a court order if there is no standard "front door" means for doing so?

So let's return to the basics. CALEA consists of statutory provisions that require telecommunication providers to support law enforcement requests for forensic evidence pursuant to court orders. Pursuant to those provisions, the DOJ via FCC establishes generic technical and administrative requirements. Industry bodies develop a standard "handover interface" for the law enforcement support requests. Individual providers or vendors decide on their own specific technical implementations for acquisition and mediation for that interface. It is as simple as that. My legal comment that the "information services" exclusion in the 1994 CALEA statute is a half-century-old anachronistic absurdity was supported by the U.S. Court of Appeals when it reviewed the FCC's application of CALEA to IP-based services, as well as its virtual disappearance from contemporary use.

Extra points are earned for noting that the telecommunication networks and marketplace today are global, that they largely consist of computer systems using software that is potentially vulnerable, and that requirements similar to CALEA exist in essentially every nation. Thus, vendors and transnational providers have long designed for global handover standards, and the collaboration occurs in international industry forums. Similarly, vendors and providers have shifted to overlay forensic systems that reduce the vulnerabilities in today's networks.

Instead of dwelling on technologies that no longer exist, we should be addressing the "What's Next?" dimension of this discussion, and noting that cyber crime is rampant and exponential, and new mobile technologies and "cloud-based virtualized" services, inter alia, are profoundly redesigning telecommunication infrastructures and offerings. These new technologies will require new approaches to keep law enforcement officials from "going dark." Law enforcement, providers, and vendors are working together globally to deal with the challenges in multiple forums, even if Congress remains in stasis. This is the real world.

CALEA: "What's Next?"—The Last Word(s)

Susan Landau

Rutkowski writes, "[Landau] argues that failure of the DOD and the State Department to publicly support DOJ CALEA arguments during congressional hearings means DOJ's proposals are 'risky.' Huh, really?"

Yes, really. Indeed, for well over a decade, the U.S. government has sought to secure civilian communications, even though doing so might impede legally authorized interceptions. In 1995, the U.S.

Naval Research Laboratory began supporting a project to hide user identities and their network activity through routing communications over a set of encrypted connections.[31] Only the endpoints would know who is communicating with whom. In 2000, the U.S. government relaxed its cryptographic export controls.[32] This meant that strong cryptography would be available in domestic products as well. In 2001, the National Institute of Standards and Technology approved the Advanced Encryption Standard (AES) for protecting electronic data.[33] (AES uses keys of 128, 192, and 256 bits, making the algorithm, even at its shortest key length, a **billion billion** times more difficult to break under a brute-force attack than the DES algorithm it replaced.) In 2003, the National Security Agency approved AES for protecting classified information;[34] broadening the market increases the availability of AES products for the civilian sector. In 2005, the NSA announced Suite B, a set of public algorithms for securing a communications network.

In 2010, the Obama administration released its *International Strategy to Secure Cyberspace*[35] based on "core commitments to *fundamental freedoms, privacy*, and the *free flow of information*,"[36] while the Department of Defense funded research to "enable safe, resilient communications over the Internet, particularly in situations in which a third party is attempting to discover the identity or location of the

---

31. *Onion Routing: Brief Selected History*, http://www.onion-router.net/History.html.

32. Dep't of Commerce, Bureau of Export Admin., 15 C.F.R. pts. 734, 740, 742, 770, 772, and 774, Docket No. RIN: 0694-AC11, Revisions to Encryption Items. Effective Jan. 14, 2000.

33. Federal Information Processing Standards Pub. 197, Announcing the Advanced Encryption Standard (Nov. 26, 2001).

34. Committee on National Security Systems, National Policy on the Use of the Advanced Encryption Standard (AES) to Protect National Security Systems and National Security Information, Policy No. 15, Fact Sheet No. 1 (Fort Meade, Md.: June 2003).

35. Office of the President, Int'l Strategy for Cyberspace: Prosperity, Security, and Openness in a Networked World (May 2011).

36. International Strategy, p. 5.

end users."[37] Such technologies secure U.S military communications, as well as those of human rights activists and journalists. Of course, these tools will also be employed domestically; their use is likely to make law enforcement investigations more difficult. But securing communications is of sufficient importance that the U.S. government is willing to make the trade-off.

In the 1990s, government encryption policy delayed the adoption of cryptography into U.S. communication and computer infrastructure.[38] The slowdown in deploying security technologies contributed substantially to U.S. industry and the U.S. government itself being more vulnerable to cyber exploitations.

Information services are harder to secure than the phone network. By creating architected security breaches ripe to be exploited, extending CALEA to information services would be the equivalent of the cryptography export control stance on steroids. No matter how much law enforcement might argue it is critical to extend CALEA to information services, we cannot afford a "solution" that will so dangerously damage our security.

---

37. Defense Advanced Research Projects Agency, *Safer Warfighter Communications*, Broad Agency Announcement, DARPA-BAA-10-69 (May 20, 2010), p. 2.

38. *See, e.g.*, Appl. Ref. No: Z066051/G006298; Philip R. Karn, Jr., Plaintiff v. U.S. Dep't of State and U.S. Dep't of Commerce, and William A. Reinsch, Undersec'y of Commerce for the Bureau of Export Admin., U.S. Dist. Ct. of App., Civ. A. No. 95-1812(LBO); Appl. Ref. No: Z066051/G006298; Lee Tien, letter to William Reinsch in reference to Appl. Ref. No.: Z066051/G006298.

Part Three:
Legal Frameworks for
Projecting Force

Chapter Eight
A Legal Framework for Targeted Killing

Amos N. Guiora

Monica Hakimi

Introduction

It is preferable to capture suspected terrorists where feasible—among other reasons, so that we can gather valuable
intelligence from them; but we must also recognize that there
are instances where our government has the clear authority—and, I would argue, the responsibility—to defend the
United States through the appropriate and lawful use of lethal force. . . . The unfortunate reality is that our nation
will likely continue to face terrorist threats that, at times,
originate with our own citizens. When such individuals take
up arms against this country and join al Qaeda in plotting
attacks designed to kill their fellow Americans, there may
be only one realistic and appropriate response.[1]
    —Attorney General Eric Holder, February 27, 2012

Holder . . . [has] argued that the Executive Branch, alone,
should determine whether te due process requirement is sat-
isfied when the government claims law of war or self-defense authority to kill. In a system of constitutional checks
and balances, that simply cannot be the case. Courts must

---

1. Eric Holder, Attorney General of the United States, Speech at Northwestern
University School of Law (Feb. 27, 2012).

have a role in determining whether the government's authority to kill its own citizens is legal and whether a decision to kill complies with the Constitution. Otherwise, the government can wield the power to take life with impunity. We should not trust any president—whether this one or the next—to make such momentous decisions fully insulated from judicial review.[2]

—ACLU statement responding to Holder

• Since 2004, the United States has conducted roughly 300 counterterrorism drone strikes in Pakistan, Afghanistan, and other areas.[3]

Targeted Killing:
Lawful if Conducted in Accordance with
the Rule of Law

Amos N. Guiora

T he drone attack that killed Anwar al-Awlaki has been the subject of innumerable articles, commentaries, and public discussion. The fact Al-Awlaki is an American citizen has dramatically increased the public scrutiny of the drone policy initiated by President George W. Bush in the aftermath of 9/11 and significantly enhanced by President Barack Obama. The discussion is healthy and essential in large part because drone warfare will play an increasingly important role in the future of operational counterterrorism.

From the perspective of the nation-state, the benefits of targeted killing are clear: aggressive measures against identified targets with

---

2. Nathan Wesler, *In Targeted Killing Speech, Holder Mischaracterizes Debate Over Judicial Review,* ACLU, March 5, 2012 (https://www.aclu.org/blog/national-security/targeted-killing-speech-holder-mischaracterizes-debate-over-judicial-review).

3. *The Year of the Drone,* COUNTERTERRORISM STRATEGY INITIATIVE, March 13, 2012 (http://counterterrorism.newamerica.net/drones).

minimal, if any, risk to soldiers for the obvious reason that the killings are conducted from an unmanned aerial vehicle. While the risks to soldiers are minimal, there are other risks that are not insignificant. Particularly, there is always the risk of collateral damage, and there are also legitimate concerns regarding how a target is defined as legitimate.

While I believe the Al-Awlaki killing lawful, I am deeply troubled by the broad rationale articulated by the Obama Administration. Yes, the Al-Awlaki killing reflects aggressive self-defense coupled with a respect for the obligation to minimize collateral damage. However, the Administration failed to articulate exactly how, beyond mere speech, Al-Awlaki was connected to terrorist activity. The mere "likelihood" of membership in a terrorist organization is highly problematic.

The essence of targeted killing, arguably the most aggressive form of operational counterterrorism, is killing an individual the nation-state has identified as posing a danger to national security, and there is no alternative, in the name of national security, but to kill the individual. The decision must reflect a rigorous application of "checks" to ensure that the decision is neither arbitrary nor in violation of international law and core principles of morality in armed conflict.

I am a firm believer in the nation-state's right to engage in aggressive, preemptive self-defense subject to powerful restraints and conditions. I advocate a measured, cautious approach to targeted killing with the understanding that the nation-state has the absolute right—and obligation—to protect its civilian population. However, that absolute right does not translate into an unlimited right.

After all, conducting operational counterterrorism divorced from a balanced approach results in violations of international law obligations, violates principles of morality in armed conflict, and results in policy ineffectiveness. The challenge in the targeted killing paradigm is to identify the specific individual deemed a legitimate target and to implement the policy in a manner reflecting respect for international law.

At its core, targeting killing reflects aggressive self-defense. Needless to say, neither the policy (in principle) nor its application (in specific) is controversy-free or immune to criticism. In the seminal case regarding targeted killings, the Israel Supreme Court sitting as the High Court of Justice held:

> The approach of customary international law applying to armed conflicts of an international nature is that civilians are protected from attacks by the army. However, that protection does not exist regarding those civilians "for such time as they take a direct part in hostilities" (§ 51(3) of *The First Protocol*). Harming such civilians, even if the result is death, is permitted, on the condition that there is no other less harmful means, and on the condition that innocent civilians nearby are not harmed. Harm to the latter must be proportionate. That proportionality is determined according to a values based test, intended to balance between the military advantage and the civilian damage. As we have seen, we cannot determine that a preventative strike is always legal, just as we cannot determine that it is always illegal. All depends upon the question whether the standards of customary international law regarding international armed conflict allow that preventative strike or not.[4]

Active self-defense (in the form of targeted killing), if properly executed, not only enables the state to more effectively protect itself within a legal context but also leads to minimizing the loss of innocent civilians caught between the terrorists (who regularly violate international law by using innocents as human shields) and the state.

Active self-defense aimed at the terrorist must contain an element of "pinpointing" so that the state will attack only those terrorists who are directly threatening society. The first step in creating an effective counterterrorism operation is analyzing the threat, including the na-

---

4. Public Committee Against Torture in Israel, Palestinian Society for the Protection of Human Rights and the Environment v. The Government of Israel, and others, HCJ 769/02, 40.

ture of the threat, who poses it, and when it is likely to be carried out. It is crucial to assess the imminence of any threat, which significantly impacts the operational and legal choices made in response.

To ensure both the legality and morality of drone strikes, I propose the following standards:

1.	A target must have made significant steps directly contributing to a planned act of terrorism.
2.	An individual cannot be a legitimate target unless intelligence action indicates involvement in future acts of terrorism.
3.	Before a hit is authorized, it must be determined that the individual is still involved and has not proactively disassociated from the original plan.
4.	The individual's contribution to the planned attack must extend beyond mere passive support.
5.	Verbal threats alone are insufficient to categorize an individual as a legitimate target.

The significant advantage of active self-defense—subject to recognized restraints of fundamental international law principles—is that the state can act against terrorists who present a real threat *prior* to the threat materializing (based on sound, reliable, and corroborated intelligence information or sufficient criminal evidence) rather than reacting to an attack that has already occurred.

While there is much disagreement among legal scholars as to the meaning (and, subsequently, timing) of words such as "planning to attack," the doctrine of active self-defense enables the state to undertake all operational measures required to protect itself.

Lawful targeted killing must be based on criteria-based decision making, which increases the probability of correctly identifying and attacking the legitimate target. The state's decision to kill a human being in the context of operational counterterrorism must be predicated on an objective determination that the "target" is, indeed, a legitimate target. Otherwise, state action is illegal, immoral, and ultimately ineffective. It goes without saying that many object to the killing of a human being when less lethal alternatives are available to neutralize the target.

Any targeted killing decision must reflect consideration of four distinct elements: law, policy, morality, and operational considerations. Traditional warfare once pitted soldier against soldier, plane against plane, tank against tank, and warship against warship.

Present and future asymmetric conflict reflects state engagement with non-state actors. In the targeted-killing paradigm, the questions—*who* is a legitimate target and *when* is the target legitimate—are at the core of the decision-making process. How both questions—in principle and practice alike—are answered determines whether the policy meets international law obligations.

The dilemma of the decision maker in the targeting paradigm is extraordinary; the time to make the decision is short, limited, and stress-filled. After all, national security is at stake. However, not all individuals identified as posing threats to national security are indeed *those* persons. A criteria-based decision-making model is necessary to ensure that the identified target is, indeed, the legitimate target.

Any use of force under international law must meet a four-part test: (1) It must be proportionate to the threat posed by the individual; (2) collateral damage must be minimal; (3) alternatives have been weighed, considered, and deemed operationally unfeasible; and (4) military necessity justifies the action. In addition, all these principles build on the fundamental international law principle of distinction, which requires that any attack distinguish between those who are fighting and those who are not in order to protect innocent life.

Regardless of whether a target is legitimate, if an attack fails to satisfy the requirements listed above, it will not be lawful. Thus, the Israeli Special Investigatory Commission[5] examining the targeted

---

5. http://www.pmo.gov.il/PMOEng/Communication/Spokesman/2011/02/spokeshchade270211.htm (last visited March 8, 2011).

killing of Saleh Shehadah concluded that although the targeting of Shehadeh—head of Hamas's Operational Branch and the driving force behind many terrorist attacks—was legitimate, the extensive collateral damage caused in the attack was disproportionate.

In any targeted killing decision, three important questions must be answered: First, can the target be identified accurately and reliably? Second, does the threat the target poses justify an attack at that moment or are there alternatives? And, finally, what is the extent of the anticipated collateral damage?

To answer these questions using the criteria-based process, extensive intelligence must be gathered and thoroughly analyzed. The Intelligence Community receives information from three different sources: human (such as individuals who live in the community about which they are providing information to an intelligence officer), signal intelligence (such as intercepted phone and e-mail conversations), and open sources (the Internet and newspapers, for example).

One of the most important questions in putting together an operational "jigsaw puzzle" is whether the received information is "actionable," that is, does the information warrant a response? This question is central to the criteria-based method, or at least to a process that seeks—in real time—to create objective standards for making decisions based on imperfect information (as almost all intelligence is). It is essential that intelligence information, particularly from humans, be subjected to rigorous analysis.

Targeted killing is a legitimate and effective form of active self-defense provided that it is conducted in accordance with clear international law principles and a narrow definition of legitimate target; otherwise, it reflects state action bound neither by the rule of law nor constraints of morality. Morality and legality demand that operational counterterrorism measures reflect criteria-based decision making, otherwise the stakes and the price are too high.

A Functional Method for Defining the Authority to Target

Monica Hakimi

I agree with much of what Professor Amos Guiora says, but I disagree with the method he uses to get there. And I believe the method matters. Guiora assesses targeting operations under an "active self-defense" paradigm, with elements from both the *jus ad bellum* (the law governing the use of force) and the *jus in bello* (the law governing the conduct of hostilities). Under Guiora's paradigm, a state may target terrorism suspects in anticipatory self-defense if: (1) targeting is proportional to their threat; (2) collateral damage is minimized; (3) alternatives to targeting are infeasible; and (4) military necessity justifies the action. Guiora does not explain why that paradigm is the correct one.

In fact, the *ad bellum* rules on defensive force probably do not govern Guiora's poster-child case—the U.S. operation targeting Anwar al-Awlaki in Yemen. The *jus ad bellum* does not constrain the use of force by one state in another state where that second state consents. Yemen appears to have consented to the operation against Al-Awlaki. Moreover, neither the *jus ad bellum* nor the traditional *jus in bello* requires a state to consider alternatives to lethal force— Guiora's third criterion—if someone is a legitimate target. Finally, though Guiora argues that someone's membership in al Qaeda is an insufficient basis for targeting him, many *in bello* experts treat membership in an organized armed group as dispositive. Rather than reflect existing law, then, Guiora's model is some kind of "hybrid." He has presented his own normative vision on when targeting should be lawful.

I assume that Guiora developed that hybrid because he believes that the traditional wartime paradigm—designed for interstate wars— is poorly suited for the fight against al Qaeda. Similarly, I assume that Guiora rejects international law's presumptive alternative—applying human rights law—because he believes that it, too, is inapposite. The human rights norms on targeting were developed for law enforcement settings. They would prohibit operations that Guiora would permit. For example, human rights law generally prohibits a state from targeting someone who is not on the verge of killing. Guiora does not require that kind of imminent threat. But Guiora does not explain why *his* model is preferable to the alternatives. Why should decision makers assess targeting operations using his four criteria, instead of applying the conventional wartime paradigm, the law enforcement paradigm of human rights law, or a hybrid advanced by someone else?

In other work, I argue that the current method for assessing targeting operations—which requires first identifying the correct legal paradigm and then applying the norms as specified for that paradigm—is misguided.[6] The method presumes that international law's different paradigms operate independently and sometimes incompatibly. But as I demonstrate, three core principles animate all the international law on targeting: the *jus in bello* for combatants, the *jus in bello* for civilians, and human rights law.

- The *liberty-security principle* identifies the outer bounds of permissible state action. The security benefits of containing someone's threat must be proportional to or outweigh the costs of life. Targeting usually satisfies that principle if the person poses an active threat of death or serious bodily injury. In that event, the security benefit of containing the threat (protecting life or limb) is proportional to the liberty cost (taking life).

- The *mitigation principle* further restricts the authority to target by requiring states to try to lessen the liberty costs. States must try to contain threats using reasonable, nonlethal alternatives to targeting—most obviously, capture and detention. Reasonableness here depends primarily on two factors. One is the level of state control. The more control

---

6. *See* Monica Hakimi, *A Functional Approach to Targeting and Detention*, 110 MICH. L. REV. __ (forthcoming 2012).

the state has, the more reasonable it will be to capture the suspect. The second factor is the relative efficacy of that alternative. States need not try to capture someone if doing so might compromise a security mission or fail to mitigate the liberty costs.

- The *mistake principle* requires states to try to verify that: (1) the person being targeted (2) poses a sufficiently serious threat (3) that cannot reasonably be contained less intrusively. In other words, the state must exercise due diligence to avoid mistakes and establish a reasonable, honest belief that its conduct is lawful. That diligence is generally less when states act in the heat of the moment than with time for deliberation. With time, states have more opportunity to ensure the accuracy of their assessments and consider the alternatives.

Those three principles govern all targeting operations but require different results depending on the facts. They may lead to results that Guiora would support. My liberty-security principle is similar to his proportionality criterion. My mitigation principle is like his criterion that states consider alternatives to targeting. And though none of his criteria specifically addresses mistakes, Guiora argues that states must gather and thoroughly analyze intelligence to ensure the accuracy of their operations. In substance, then, we seem to agree on quite a bit—at least at this level of generality. (My method does not address the permissible collateral damage. I agree with Guiora that, consistent with both the *jus in bello* and human rights law, any collateral damage must be minimized.)

But methodologically, we differ. I argue for assessing *all* targeting operations by reference to the above three principles. Most international lawyers invoke their preferred legal paradigm—Guiora selects a hybrid for "active self-defense"—and then apply the norms associated with that paradigm. As I demonstrate in my other work, that latter method breeds uncertainty and undermines the discursive process by which the law might adapt to modern challenges or hold decision makers accountable.[7] It breeds uncertainty because decision makers sometimes disagree on the governing paradigm even when they agree in substance. Agreeing on the paradigm for one case may threaten slippery-slope implications for other cases. Or it may require a hybrid that, like Guiora's, is not accessible under existing law. Debating the applicable paradigm obscures areas of substantive agreement.

More importantly, my method would focus decision makers on the considerations that actually drive legal outcomes. Decision makers now justify particular outcomes by invoking their preferred paradigms. Those who disagree on the applicable paradigm talk past each other, applying different norms to assess the same or similar conduct. That enfeebled discourse is a problem because international law—and especially the law on targeting—primarily operates discursively. When the legal process works well, it provides a common language with which decision makers may justify their positions and respond to counterarguments. Eventually, they may converge on particular outcomes and resolve substantive uncertainties. But even when they disagree on substance, the discursive process helps constrain their discretion. The more persuasively an actor defends its position, the less pressure it confronts to alter its conduct. Conversely, the more compelling the counterarguments, the more it must change its behavior or refine its position to avoid condemnation.

Consider the Al-Awlaki case. The U.S. government justifies that and similar operations by invoking a global armed conflict against al Qaeda. It claims that the *jus in bello* for traditional combatants also governs operations against members of al Qaeda. The United States makes that claim, even though it does not intend to target al Qaeda members worldwide. Rather, it claims a global conflict, because it views the presumptive alternative—applying human rights law wherever U.S.–al Qaeda hostilities are *not* active—as sometimes too limiting. Hybrids such as Guiora's might be normatively appealing but are not now grounded in existing law. Thus, the method for assessing

---

7. *Id.*

targeting operations pushes the United States toward a legal position that is more extreme than its practice. Meanwhile, those who protest the U.S. practice lack effective tools for holding it accountable. Just as the wartime paradigm is ineffective in legitimizing U.S. operations, the law enforcement paradigm is ineffective in constraining those operations. The United States easily dismisses human rights law as inapplicable. Of course, the Al-Awlaki operation fits neatly into none of the existing paradigms. But because the current method requires identifying the correct paradigm before assessing state conduct, decision makers endlessly debate which of those ill-fitting options is preferable.

By contrast, my method invites the United States to defend its operations on the merits—by reference to the principles that animate all existing law. Compromise positions may satisfy U.S. security needs while better legitimizing its operations internationally. The United States clearly seeks to do both. Here is President Obama's chief counterterrorism adviser, John Brennan:

> The effectiveness of our counterterrorism activities depends on the assistance and cooperation of our allies. . . . But their participation must be consistent with their laws, including their interpretation of international law. . . . The more our views and our allies' views on these questions converge, without constraining our flexibility, the safer we will be as a country.[8]

Whereas the current method pushes the United States toward the extreme armed-conflict claim, mine would encourage more moderation—which the United States itself seeks. Over time, the United States and other actors may narrow their disagreements and resolve when states may target terrorism suspects extraterritorially. For example, although the United States and Human Rights Watch disagree

---

8. John O. Brennan, Assistant to the President for Homeland Security and Counterterrorism, *Strengthening our Security by Adhering to our Values and Laws*, Speech at the Harvard Law School Program on Law and Security (Sept. 16, 2011).

on the applicable legal paradigm, they seem to agree that the Al-Awlaki operation was permissible in part because capture was infeasible. Yet even where substantive disagreements persist, my method would better hold the United States accountable. It would require the United States to defend its conduct on the merits, instead of by reference to opaque legal paradigms. Some positions—such as the claim that it may target all al Qaeda members—would be considerably more difficult to defend.

Targeted Killing: Reply to Monica Hakimi

Amos N. Guiora

To the naked eye, Professor Monica Hakimi and I agree: Targeted killing is lawful provided it is subject to criteria and standards. Perhaps we reach this conclusion from different perspectives and distinct analysis, but the conclusion is similar. In other words, targeted killing is legal; the question is under what conditions. A close reading of Professor Hakimi's thoughtful and well-written response to my initial essay suggests concern with my analysis of imminence; in other words, how do we determine whether the threat posed is sufficiently imminent to determine that the potential target is, indeed, a legitimate target.

Professor Hakimi is spot-on in highlighting this issue. Similarly, she is correct in suggesting that my essay proposes a rearticulation of international law to account for a new operational model. There is, frankly, discomfort in proposing new models; ad hoc solutions are inherently dangerous because their limits are unclear. In that vein, as history continuously suggests, unlimited executive power in the face of threats raises deeply important questions and concerns.

That said, to apply traditional models to new threats is similarly problematic; the challenge is implementing proactive operational measures subject to rigorous checks and balances with narrow definitions of critical terms. As much discussed in scholarly literature on war and international law—and as Professor Hakimi correctly notes—the term *imminence* is elusive, problematic, and subject to wide interpretation. Imminence, in the targeted killing paradigm, suggests that unless the nation-state decisively engages a particular individual deemed to pose a direct threat, then innocent civilians will be harmed.

For example, to successfully conduct a suicide bombing requires a doer (the bomber), a sender (responsible for the operation in all parameters), a logistician (responsible for all operational logistics), and a financier (responsible for financing the attack, whether directly or indirectly). All four actors are essential, individually and collectively.

The proactive self-defense model at the core of targeted killing requires determining when each actor is a legitimate target predicated on an imminence analysis. Too broad a definition violates international law and morality in armed conflict standards; too narrow a definition unnecessarily endangers innocent civilians to whom the nation-state owes a duty to protect. Based on international law principles of military necessity and proportionality, along with the requirement to minimize collateral damage and to pursue alternatives, the four actors are legitimate targets at distinct times.

The doer is a legitimate target when about to commit a suicide bombing; the sender is a legitimate target 24/7 regardless of specific actions at the moment provided collateral damage is minimized; the logistician is a legitimate target when involved in planning an attack, with the understanding that continued involvement poses a greater threat to national security than the doer of a specific attack; the financier, while largely an unresolved dilemma, is a legitimate target more akin to the sender than to the logistician and immeasurably more so than the doer. After all, financiers are to terrorism what intelligence information is to counterterrorism. There is no terrorism without financiers and there is no counterterrorism without intelligence information.

Where, then, does this leave us with respect to the questions Professor Hakimi posed? While recommending new paradigms is inevitably a risky proposition, the core question is whether the nation-state has the requisite tools to effectively engage in aggressive self-defense against an amorphous target. Professor Hakimi and I agree that an overbroad definition of a legitimate target is a danger-ous road to travel. Similarly, we agree that standardless targeted kill-ing models not predicated on well-defined criteria pose an extraordi-nary danger to the rule of law and morality standards. Nevertheless, while debate is important—particularly given the dangers inherent to excessive state power—it is important to cut to the chase.

To that end, the working model proposed above for defining both the legitimate target categories and when those targets may be legitimately engaged suggests a way forward. While inevitably sub-ject to criticism and concern, it reflects a balancing approach re-quired by international law in a conflict that I have previously re-ferred to as "mission impossible." After all, identifying a legitimate target in the traditional war paradigm posed minimal challenges to operational decision makers; defining a legitimate target in the state/ non-state paradigm poses extraordinary challenges. Targeted killing is the most aggressive form of self-defense; in the present paradigm, its morality, legality, and effectiveness demand narrow definitions of legitimate target strictly applied. That is the model I have proposed. How criteria-based decision making is applied determines whether the nation-state conducts itself in accordance with international law.

Targeted Killing: Reply to Amos Guiora

Monica Hakimi

In our first round of debate, Professor Guiora proposed assessing targeting operations using a model of "active self-defense." I pre-sented a counterproposal. I argued that certain basic principles—which I term liberty-security, mitigation, and mistake—determine when targeting is lawful.

Substantively, our proposals have much in common. They also have some differences. For example, Guiora asserts that states may target people who only finance terrorism. My liberty-security prin-ciple probably prohibits that result. Financiers do not themselves threaten bodily integrity and are too removed from the harm to jus-tify targeting them.

No matter how one assesses those substantive differences, my proposal is preferable to Guiora's because mine is methodologically sound. First, my proposal is rooted in existing law. Its three principles drive settled outcomes under both the *jus in bello* and human rights law. In contrast, Guiora's model reflects his own normative vision. Second, my principles apply in all contexts. Guiora does not identify precisely when his model applies, but he apparently intends for it to apply in only *some* contexts. He did not contest my suggestion that, before using his model, decision makers would have to determine whether it even applies. Third and as explained in my initial response, my proposal would invigorate international law's discursive process. Rather than debate which targeting model applies, decision makers would focus on the considerations that actually drive legal outcomes. That substantive discourse helps develop the law and hold decision makers accountable.

Thus far, our debate has focused on the *in bello* and human rights restraints on targeting. Targeting sometimes also implicates the *jus ad bellum*. The *jus ad bellum* regulates when states may use force against non-state actors in other states. Such force is lawful when the territorial state consents. It probably also is lawful when the territorial state is unable or unwilling to contain the non-state threat.

Since the September 11 attacks, states have more frequently used force to incapacitate terrorists in other states—either with consent or under the unable-or-unwilling standard. Those operations may be lawful under the *jus ad bellum* without falling squarely in any *in bello* or human rights paradigm. For example, one-off operations might not cause sufficient violence to trigger an armed conflict. The *jus in bello* would not apply. Similarly, human rights law might not apply. The extent to which it applies extraterritorially is contested and uncertain. Thus, some lawyers suggest that the *ad bellum* license to use force effectively displaces any *in bello* or human rights restraints. I disagree (and I presume from Guiora's comments that he would disagree, as well). The *jus ad bellum* is concerned primarily with protecting state sovereignty. The *jus in bello* and human rights law are about protecting individuals. Both interests are at stake when states use force extraterritorially.

In my view, decision makers must adapt the *in bello* and human rights protections to account for developments under the *jus ad bellum*. My proposal enables that move. Recall that my mitigation principle requires states to pursue reasonable, nonlethal alternatives to targeting. Reasonableness here depends on two factors. One is state control. The greater a state's control, the more varied its toolbox, and the more comfortably it may contain a threat without resorting to deadly force. The second factor is the relative efficacy of an alternative. States need not pursue measures that are unsuitable for or realistically might compromise the security mission.

The fact that an operation is lawful under the *jus ad bellum* does not make it so under my proposal. Capture might be reasonable—and therefore required—even if the territorial state consents to military force. States usually must cooperate to apprehend terrorists with the tools of law enforcement. Yet the circumstances that justify nonconsensual force might also justify taking human life. Capture might be unreasonable when the territorial state is unable or unwilling.

Consider the U.S. operation against Osama Bin Laden. Pakistan did not consent to that operation. The *jus ad bellum* asks whether Pakistan was unable or unwilling to incapacitate Bin Laden—for example, because of incompetence or corruption. If Pakistan was unable or unwilling, then working with its law enforcement apparatus was almost certainly unreasonable for purposes of the mitigation principle. To be sure, the United States might have had other alternatives for containing Bin Laden's threat. But the mitigation principle suggests that, if the territorial state is uncooperative, the targeting state should have more than its ordinary, law enforcement authorities. The United States had considerably fewer tools for controlling the situation in Abbottabad than it has domestically.

Chapter Nine
Cyberwar

Stewart Baker

Charles J. Dunlap Jr.

Introduction

In June of 1982, the North American Aerospace Defense Command (NORAD) registered a 3-kiloton explosion in the middle of Siberia. While NORAD's instruments indicated a nuclear explosion, no electromagnetic pulse followed. Also, it was known that no missile silos were housed at the detonation site, which was uninhabited. The Soviets had stolen turbine and pressure-regulation software from a Canadian technology firm to manage their new Trans-Siberian pipeline. They didn't realize that the CIA had become aware of the initiative and, according to published accounts, reengineered the software to go haywire.[1]

This act was one of the earliest uses of computer programming to physically damage enemy infrastructure. Today, one can mount such an attack even when the targeted system is not connected to the Internet. Such was the case with the Stuxnet virus that infected computers regulating Iran's nuclear centrifuges.[2]

---

1. William Safire, *The Farewell Dossier*, N.Y. Times, Feb. 2, 2004, *available at* http://www.nytimes.com/2004/02/02/opinion/the-farewell-dossier.html.
2. Maziar Bahari, Ronen Bergman & John Barry, *The Shadow War*, The Daily Beast, *available at* http://www.thedailybeast.com/newsweek/2010/12/13/the-covert-war-against-iran-s-nuclear-program.html.

We live in a world that is increasingly run and regulated using computer systems and the Internet. This has grown to include most financial commerce, utilities, communications, and military technology. As we become more reliant on cyberspace, the prospective damage inflicted by a cyber attack increases.

During Desert Storm in 1990, the United States demonstrated what happens when two armies meet and one possesses cyber capabilities that the other lacks. However, U.S. cyber superiority in Desert Storm was primarily a reflection of advanced communications and targeting systems. Today, cyber warfare has morphed into a new era where offensive cyber attacks may target the national security computer systems of the enemy and can be coordinated with kinetic maneuvers. The first-ever exercise of this tactic occurred in August of 2008, when Russian forces invaded Georgia in tandem with a series of cyber attacks affecting Georgia's infrastructure. The attacks were successfully debilitating, blocking Internet access and disabling communications (they also targeted banking, media, and government websites).[3]

Ironically, because Georgia's cyber infrastructure was relatively underdeveloped, the attacks were not so crippling as they might have been against a more cyber-reliant country. The ramifications of an attack against this type of target became evident a year *before* the invasion of Georgia in a series of anonymous cyber attacks on Estonia in April and May of 2007—the first-ever example of a cyber attack targeting the security infrastructure of a country. Estonia relies heavily on cyber infrastructure for its voting, education, security, and financial commerce (including 95 percent of its banking transactions).[4] While the attack on Estonia was relatively unsophisticated, it demonstrated the frightening, almost limitless potential of what could happen if a country with an Internet-centric infrastructure suffers an advanced, coordinated cyber attack.

---

3. William C. Ashmore, *Impact of Alleged Russian Cyber Attacks*, Baltic Security and Defense Review, Vol. 11, 2009, *available at* http://jmw.typepad.com/files/ashmore—impact-of-alleged-russian-cyber-attacks-.pdf.

4. *Id.*

In June of 2009, President Barack Obama created CYBERCOM, a battle command tasked exclusively with coordinating cyber operations and security for the armed forces.[5]

In the kinetic realm, the U.S. military and Intelligence Community (IC) have recently experienced increased convergence and overlap between "Traditional Military Activities" (TMA) and "Covert" intelligence operations (as defined by Titles 10 and 50 of the U.S. legal code). There remain countless unresolved policy questions governing when and how authorities may respond to cyber attacks.

Law and Cyberwar—The Lessons of History

Stewart Baker

Lawyers don't win wars.

But can they lose a war? We're likely to find out, and soon. Lawyers across the government have raised so many show-stopping legal questions about cyberwar that they've left our military unable to fight, or even plan for, a war in cyberspace.

No one seriously denies that cyberwar is coming. Russia may have pioneered cyber attacks in its conflicts with Georgia and Estonia, but cyber weapons went mainstream when the developers of Stuxnet sabotaged Iran's Natanz enrichment plant, proving that computer network attacks can be more effective than 500-pound bombs. In war, weapons that work get used again.

Unfortunately, it turns out that cyber weapons may work best against civilians. The necessities of modern life—pipelines, power grids, refineries, sewer and water lines—all run on the same industrial control systems that Stuxnet subverted so successfully. These systems may be even easier to sabotage than the notoriously porous

---

5. National Security Council, *Cybersecurity Progress After President Obama's Address*, The White House, July 14, 2010, *available at* http://www.whitehouse.gov/administration/eop/nsc/cybersecurity/progressreports/july2010.

computer networks that support our financial and telecommunications infrastructure.

No one has good defenses against such attacks. The hackers will get through.

Even very sophisticated network defenders—RSA, HBGary, even the Department of Defense (DOD) classified systems—have failed to keep attackers out. Once they're in, attackers have stolen the networks' most precious secrets. But they could just as easily bring the network down, possibly causing severe physical damage, as in the case of Stuxnet.

So as things now stand, a serious cyber attack could leave civilians without power, without gasoline, without banks or telecommunications or water—perhaps for weeks or months. If the crisis drags on, deaths will multiply, first in hospitals and nursing homes, then in cities and on the road as civil order breaks down. It will be a nightmare. And especially for the United States, which has trusted more of its infrastructure to digital systems than most other countries have.

We've been in this spot before. As General William Mitchell predicted, airpower allowed a devastating and unprecedented strike on our ships in Pearl Harbor. We responded with an outpouring of new technologies, new weapons, and new strategies.

Today, the threat of new cyber weapons is just as real, but we have responded with an outpouring, not of technology or strategy but of law review articles, legal opinions, and legal restrictions. Military lawyers are tying themselves in knots trying to articulate when a cyber attack can be classified as an armed attack that permits the use of force in response.[6] State Department and National Security Council lawyers are implementing an international cyberwar strategy that relies on international law "norms" to restrict cyberwar.[7] CIA law-

---

6. *See* WILLIAM A. OWENS ET AL., TECHNOLOGY, POLICY, LAW, AND ETHICS REGARDING U.S. ACQUISITION AND USE OF CYBERATTACK CAPABILITIES 239–92 (2009), *available at* http://www.nap.edu/catalog.php?record_id=12651.

7. THE WHITE HOUSE, INTERNATIONAL STRATEGY FOR CYBERSPACE: PROSPERITY, SECURITY, AND OPENESS IN A NETWORKED WORLD 9 (2011), *available at* http://www.whitehouse.gov/sites/default/files/rss_viewer/international_strategy_for_cyberspace.pdf.

yers are invoking the strict laws that govern covert action to prevent the Pentagon from launching cyber attacks.[8] Justice Department lawyers are telling our military that it violates the law of war to do what every cyber criminal has learned to do—cover their tracks by routing attacks through computers located in other countries.[9] And the Air Force recently surrendered to its own lawyers, allowing them to order that all cyber weapons be reviewed for "legality under [the law of armed conflict], domestic law and international law" before cyberwar capabilities are even acquired.[10] (And that's just the lawyers' first bite at the apple; the directive requires yet another legal review before the weapons are used.)[11]

The result is predictable, and depressing. Top Defense Department officials recently adopted a cyberwar strategy that simply omitted any plan for conducting offensive operations.[12] Apparently, they're still waiting for all these lawyers to agree on what kind of offensive operations the military is allowed to mount.

*   *   *

I have no doubt that the lawyers think they're doing the right thing. Cyberwar will be terrible. If the law of war can stave off the worst civilian harms, they'd argue, surely we should embrace it.

There's just one problem: That's exactly what we tried when airpower transformed war.

And we failed.

---

8. Ellen Nakashima, *Dismantling of Saudi-CIA Web Site Illustrates Need for Clearer Cyberwar Policies*, WASH. POST, March 19, 2010, *available at* http://www.washingtonpost.com/wp-dyn/content/article/2010/03/18/AR2010031805464.html.

9. Ellen Nakashima, *Pentagon Is Debating Cyber-Attacks*, WASH. POST, Nov. 6, 2010, *available at* http://www.washingtonpost.com/wp-dyn/content/article/2010/11/05/AR2010110507464.html.

10. *See* Secretary of the Air Force, Air Force Instruction 51-402, *Legal Reviews of Weapons and Cyber Capabilities* (July 27, 2011), *available at* http://www.fas.org/irp/doddir/usaf/afi51-402.pdf.

11. *Id.* 3.3.

12. *See* DEP'T OF DEFENSE, DEPARTMENT OF DEFENSE STRATEGY FOR OPERATING IN CYBERSPACE (2011), *available at* http://www.defense.gov/home/features/2011/0411_cyberstrategy/docs/DoD_Strategy_for_Operating_in_Cyberspace_July_2011.pdf.

In the first half of the 20th century, the airplane did for warfighters what information technology has done in the last quarter of a century. Like cyber attacks, airpower was first used to gather intelligence and not to fight. Perhaps for this reason, there was never a taboo about using either airpower or cyber weapons. By the time officials realized just how ugly these weapons could be, the cat was already out of the bag.

By the 1930s, though, everyone saw that aerial bombing would reduce cities to rubble in the next war. We have trouble today imagining how unprecedented and terrible airpower must have seemed at this time. Just a few years earlier, the hellish slaughter where armies met in the trenches of World War I had destroyed the Victorian world; now airpower promised to bring that hellish slaughter to the home front.

Former Prime Minister Stanley Baldwin summed up Britain's strategic position in 1932 with a candor no American leader has dared to match in talking about cyberwar:

I think it is well also for the man in the street to realize that there is no power on earth that can protect him from being bombed, whatever people may tell him. The bomber will always get through. . . . The only defence is in offence, which means that you have got to kill more women and children more quickly than the enemy if you want to save yourselves.[13]

The British may have been realists about air war, but Americans still hoped to head off the nightmare. The American tool of choice was international law. (Some things never change.) When war broke out on September 1, 1939, President Franklin D. Roosevelt sent a cable to all the combatants seeking express limits on the use of airpower and expressing his view that:

[R]uthless bombing from the air of civilians in unfortified centers of population . . . has sickened the hearts of every civilized man and woman, and has profoundly shocked the conscience of humanity. . . . I am therefore addressing this urgent appeal to every government which may be engaged

---

13. ELIZABETH M. KNOWLES, THE OXFORD DICTIONARY OF QUOTATIONS 49 (1999).

in hostilities publicly to affirm its determination that its armed forces shall in no event, and under no circumstances, undertake the bombardment from the air of civilian populations or of unfortified cities. . . .[14]

President Roosevelt had a pretty good legal case. The Hague Conventions on the Law of War, adopted just two years after the Wright Brothers' first flight, declared that in bombardments, "all necessary steps should be taken to spare as far as possible edifices devoted to religion, art, science, and charity, hospitals, and places where the sick and wounded are collected, provided they are not used at the same time for military purposes."[15] The League of Nations had recently declared that, in air war, "the intentional bombing of civilian populations is illegal."[16]

But FDR didn't rely just on law. He asked for a public pledge that would bind all sides.[17] Remarkably, he got it. The horror of aerial bombardment ran so deep in that era that England, France, Germany, and Poland all agreed—before nightfall on the same day.[18]

What's more, they tried to honor their pledges. In a June 1940 order for Luftwaffe operations against Britain, Hermann Göring "stressed that every effort should be made to avoid unnecessary loss of life amongst the civilian population."[19]

---

14. U.S. Dep't of State, Foreign Relations of the United States Diplomatic Papers, 1939 General 541–42 (1939).

15. Convention With Respect to the Laws on War on Land (Hague II), Art. 27, July 29, 1899, 32 Stat. 1803, 1 Bevans 247, *available at* http://www.icrc.org/ihl.nsf/FULL/150?OpenDocument.

16. Protection of Civilian Populations Against Bombing From the Air in Case of War, Resolution of the League of Nations, League of Nations O.J. Spec. Supp. 182, at 16 (1938), *available at* http://www.dannen.com/decision/int-law.html#D.

17. Foreign Relations of the United States at 541–42.

18. *See id.* at 542–48 (containing responses to Roosevelt's cables from the United Kingdom, France, Germany, Poland, and Italy); *see also* Fritz Kalshoven, *Bombardment, From "Brussels 1874" to "Sarajevo 2003," in* Reflections on the Law of War: Collected Essays 431, 438–40 (2007) (all sides pledged not to bomb civilians "on the basis of reciprocity").

19. Derek Wood & Derek Dempster, The Narrow Margin: The Battle of Britain and the Rise of Air Power 117 (2003).

It began to look like a great victory for the international law of war. All sides had stared into the pit of horrors that civilian bombing would open up. And all had stepped back.

It was exactly what the lawyers and diplomats now dealing with cyberwar hope to achieve.

But as we know, that's not how this story ends. On the night of August 24, a Luftwaffe air group made a fateful navigational error. Aiming for oil terminals along the Thames, they miscalculated, instead dropping their bombs in the civilian heart of the city of London.

It was a mistake. But that's not how Prime Minister Winston Churchill saw it. He insisted on immediate retaliation. The next night, British bombers hit targets in Berlin for the first time. The military effect was negligible, but the political impact was profound. Göring had promised that the Luftwaffe would never allow a successful attack on Berlin. The Nazi regime was humiliated, the German people enraged. Ten days later, Hitler told a wildly cheering crowd that he had ordered the bombing of London: "Since they attack our cities, we will extirpate theirs."[20]

The Blitz was on.

In the end, London survived. But the extirpation of enemy cities became a permanent part of both sides' strategy. No longer an illegal horror to be avoided at all costs, the destruction of enemy cities became deliberate policy. Later in the war, British strategists would launch aerial attacks with the avowed aim of causing "the destruction of German cities, the killing of German workers, . . . the disruption of civilized life throughout Germany . . . the creation of a refugee problem on an unprecedented scale, and the breakdown of morale both at home and at the battle fronts."[21]

The Hague Conventions, the League of Nations resolution, even the explicit pledges given to President Roosevelt—all these "norms"

---

20. MICHAEL KORDA, WITH WINGS LIKE EAGLES: THE UNTOLD STORY OF THE BATTLE OF BRITAIN 247 (2010).

21. BRERETON GREENHOUS, THE CRUCIBLE OF WAR, 1939-1945 725 (1994).

for the use of airpower had been swept away by the logic of the technology and the predictable psychology of war.

*　　*　　*

So, why do today's lawyers think that *their* limits on cyberwar will fare better than FDR's limits on air war?

It beats me. If anything, they have a much harder task. Roosevelt could count on a shared European horror at the aerial destruction of cities. He used that to extract an explicit and reciprocal understanding from both sides as the war was beginning. We have no such understanding, indeed no such shared horror. Quite the contrary, for some of our potential adversaries, cyber weapons are uniquely asymmetric—a horror for us, another day in the field for them. It doesn't take a high-tech infrastructure to maintain an army that is ready in a pinch to live on grass.

What's more, cheating is easy and strategically profitable. American compliance will be enforced by all those lawyers. Our adversaries can ignore the rules and say—hell, they *are* saying—"We're not carrying out cyber attacks. We're victims, too. Maybe you're the attacker. Or maybe it's Anonymous. Where's your proof?"

Even if all sides were genuinely committed to limiting cyberwar, as all sides were in 1939, we've seen that the logic of airpower eventually drove all sides to the horror they had originally recoiled from. Each side felt that it had observed the limits longer than the other. Each had lawyerly justifications for what it did, and neither understood or gave credence to the other's justifications. In that climate, all it took was a single error to break the legal limits irreparably.

And error was inevitable. Bombs dropped by desperate pilots under fire go astray. But so do cyber weapons. Stuxnet infected thousands of networks as it searched blindly for Natanz. The infections lasted far longer than intended. Should we expect fewer errors from code drafted in the heat of battle and flung at hazard toward the enemy?

Of course not. But the lesson for the lawyers and the diplomats is stark: Their effort to impose limits on cyberwar is almost certainly doomed.

No one can welcome this conclusion, at least not in the United States. We have advantages in traditional war that we lack in cyberwar. We are not used to the idea that launching even small wars on distant continents may cause death and suffering here at home. That is what drives the lawyers. They hope to maintain the old world. But they're driving down a dead end.

If we want to defend against the horrors of cyberwar, we need first to face them, with the candor of a Stanley Baldwin. Then we need to charge our military strategists, not our lawyers, with constructing a cyberwar strategy for the world we live in, not the world we'd like to live in.

That strategy needs both an offense and a defense. The offense must be powerful enough to deter every adversary with something to lose in cyberspace, and so it must include a way to identify our attacker with certainty. The defense, too, must be realistic, making successful cyber attacks more difficult and less effective because we have built resilience and redundancy into our infrastructure.

Once we have a strategy for winning a cyberwar, we can ask the lawyers for their thoughts. We can't do it the other way around.

In 1941, the British sent their most modern battleship, the *Prince of Wales*, to Southeast Asia to deter a Japanese attack on Singapore. For 150 years, having the largest and most modern navy was all that was needed to project British power around the globe. Like the American lawyers who now oversee defense and intelligence, British admirals preferred to believe that the world had not changed. It took Japanese planes 10 minutes to put an end to their fantasy, to the *Prince of Wales,* and to hundreds of brave sailors' lives.

We should not wait for our own *Prince of Wales* moment.

Lawless Cyberwar? Not If You Want to Win

Charles J. Dunlap Jr.

Lawlessness cannot win America's 21st-century wars, but it can surely lose them.

Military professionals keenly understand this verity, to include especially those responsible for conducting cyber operations. Oddly, some civilians think otherwise.

Stewart Baker, a highly respected former official of both the Department of Homeland Security and the National Security Agency, and now a member of the prestigious Washington law firm of Steptoe & Johnson, has written a lively polemic about America's presumed cyber woes. In it he claims that in their insistence upon adherence to domestic and international law, lawyers—and especially *military* lawyers—are hobbling America's cyberwar strategy.

According to Baker, lawyers "have raised so many show-stopping legal questions about cyberwar that they've left the military unable to fight, or even plan for, a war in cyberspace."

Really? If that is so, why is it that military commanders—the ones actually responsible for cyberwar fighting and planning—do not see it that way? Why are uniformed leaders expressing satisfaction with the legal framework for cyberwar, even with respect to the sensitive matter of *offensive* cyberwar that Baker seems to think the United States has forgone because of the machinations of lawyers?

Consider this: In November of 2011, Reuters reported that the commander of the U.S. Strategic Command (the parent organization of U.S. Cyber Command) acknowledged that the "U.S. military now has a legal framework to cover offensive operations in cyberspace." The officer, Air Force General C. Robert Kehler, said unequivocally that he did "not believe that we need new explicit authorities to conduct offensive operations of any kind," and added—definitively— that he did "not think there is any issue about authority to conduct operations."[22]

---

22. Jim Wolf, *U.S. Military better prepared for cyber warfare: general*, REUTERS (Nov. 16, 2011), http://www.reuters.com/article/2011/11/17/us-usa-cyber-military-idUSTRE7AG03U20111117?feedType=RSS&feedName=everything&virtualBrandChannel=11563.

23. U.S. DEP'T OF DEFENSE, DEPARTMENT OF DEFENSE CYBERSPACE POLICY REPORT: A REPORT TO CONGRESS PURSUANT TO THE NATIONAL DEFENSE AUTHORIZATION ACT FOR FISCAL YEAR 2011, Section 934 (November 2011), *available at* http://www.defense.gov/home/features/2011/0411_cyberstrategy/docs/NDAA%20Section%20934%20Report_For%20webpage.pdf.

Furthermore, the 2011 Department of Defense (DOD) *Cyberspace Policy Report* makes it clear that warfighters are ready to wage offensive cyberwar, and will do so in compliance with the existing law of armed conflict. Specifically, the DOD says:

> [DOD] has the capability to conduct offensive operations in cyberspace to defend our Nation, Allies and interests. If directed by the President, DOD will conduct *offensive cyber operations in a manner consistent with the policy principles and legal regimes that the Department follows for kinetic capabilities, including the law of armed conflict.*[23] (Italics added.)

Nevertheless, Baker argues that the cyber "offense must be powerful enough to deter every adversary with something to lose in cyberspace," and implies that America cannot do that unless it jettisons efforts to observe to the law. Curiously, he offers little evidence of any insufficiency in U.S. cyber offensive potential.

In truth, who is to say that existing U.S. offensive cyber capabilities, notwithstanding that they follow the law, are not more powerful than those of any potential adversary?

More precisely, what adversary would assume the United States is deficient in this regard? What adversaries do know is that the U.S. military is the most powerful in the world, even though it always seeks to follow the law. Why would an adversary think that U.S. cyber weapons are not as devastating as those the American military operates in every other domain?

Baker seems uncomfortable with ambiguity about U.S. cyber capabilities. In a way, that is understandable, given the dearth of information about them, but from a military perspective, making the enemy wonder about what fate might befall him were he to launch a cyber strike is a good thing. To those in the armed forces, ambiguity—especially in the cyber realm—has real deterrence value.[24]

---

24. Lt. Col. John A. Mowchan, U.S. Army, *Don't Draw the (Red) Line,* PROCEEDINGS Magazine (October 2011, Vol. 137), *available at* http://www.usni.org/magazines/proceedings/2011-10/dont-draw-red-line.

Moreover, uniformed professionals typically do not analyze one military capability in isolation from others. This is why, for example, the DOD *Cyberspace Policy Report* makes it clear that in the event of a cyber attack, the "response options may include using cyber *and/or kinetic capabilities*." (Italics added.)

In other words, America's cyber deterrence does not depend on any particular cyber capability, but includes the fearsome kinetic weaponry of the U.S. armed forces. What adversary today wants to take on America's vast arsenal of diverse military capabilities?

As the DOD report makes clear, U.S. cyber warriors are ready to wage war within the existing limits of the law of armed conflict. Baker nonetheless indicates that in his view doing so will attempt to "impose limits on cyberwar" and is, therefore, "doomed." Consequently, he implies that there should not be any limits on the way the United States wages cyberwar.

This raises an important question: Should America wage war—cyber or otherwise—without legal "limits"?

Military commanders have seen the no-legal-limits movie before and they do not like it. In the aftermath of 9/11, civilian lawyers moved in exactly that direction. Former Attorney General Alberto Gonzales, for example, rejected parts of the Geneva Conventions as "quaint." He then aligned himself with other civilian government lawyers who seemed to believe that the President's war-making power knew virtually no limits. The most egregious example of this mindset was their endorsement of interrogation techniques now widely labeled as torture.[25]

The results of the no-legal-limits approach were disastrous. The ill-conceived civilian-sourced interrogation, detention, and military tribunal policies, implemented over the persistent objections of America's military lawyers, caused an international uproar that profoundly injured critical relations with indispensable allies.[26] Even

---

25.   Gen. Charles Dunlap, *A Tale of Two Judges : A Judge Advocate's Reflections on Judge Gonzales's Apologia,* DUKE LAW FACULTY SCHOLARSHIP (2010), *available at* http://scholarship.law.duke.edu/faculty_scholarship/2274.

26.   Scott Horton, *Bush Assails the JAG Corps,* HARPER'S MAGAZINE (Dec. 16, 2007), *available at* http://harpers.org/archive/2007/12/hbc-90001929.

more damaging, they put the armed forces on the road to Abu Ghraib, a catastrophic explosion of criminality that produced what military leaders like then U.S. commander in Iraq Lieutenant General Ricardo Sanchez labeled as a "clear defeat."[27]

Infused with illegalities, Abu Ghraib became the greatest reversal America has suffered since 9/11. In fact, in purely military terms, it continues to hobble counterterrorism efforts. General David Petraeus observed that "Abu Ghraib and other situations like that are non-biodegradable. They don't go away." "The enemy," Petraeus says, "continues to beat you with them like a stick."[28] In short, military commanders want to adhere to the law because they have hard experience with the consequences of failing to do so.

Why, then, is Baker—and others—so troubled? Actually, there *are* legitimate concerns about America's cyber capabilities, but the attack on the issues is misdirected. Indeed, if Baker substitutes the term *policy maker* for *lawyer* and the term *policy* for *law,* he might be closer to the truth in terms of today's cyberwar challenges. To those with intimate knowledge of the intricacies of cyber war, it is not the "law," per se, that represents the most daunting issue; to them, it is *policy*.

For example, retired Air Force General Michael Hayden, the former head of the National Security Agency (NSA), and later Director of the CIA, told Congress in October of 2011 that America's cyber defenses were being undermined because cyber information was "horribly overclassified."[29] That issue is not sourced in lawyers, but in policy makers who could solve the classification problem virtually overnight if they wanted to.

---

27. Tom Brokaw, "Gen. Sanchez: Abu Ghraib 'clearly a defeat.'" NBC NEWS (June 30, 2004), *available at* http://msnbc.msn.com/id/5333895/#.TyGxO_nW4oh.

28. Joseph Berger, *U.S. Commander Describes Marja Battle as First Salvo in Campaign,* N.Y. TIMES, Feb. 21, 2010, *available at* http://www.nytimes.com/2010/02/22/world/asia/22petraeus.html.

29. Eric Engleman, *Chinese Cyberspying Declared 'Intolerable" by U.S. Lawmaker,* BLOOMBERG, Oct. 4, 2011, *available at* http://mobile.bloomberg.com/news/2011-10-04/chinese-cyberspying-at-intolerable-level-u-s-lawmaker-says.

That same month, General Keith B. Alexander, Commander of U.S. Cyber Command and current NSA Director, said that rules of engagement were being developed that would "help to define conditions in which the military can go on the offensive against cyber threats and what specific actions it can take." General Alexander readily acknowledges the applicability of the law of armed conflict, but suggests that challenges exist in discerning the facts and circumstances to apply to the law.[30]

This gets to the "act of war" question Baker complains about. The law does provide a framework;[31] it is up to decision makers to discern the facts to apply to that framework. Hard to do? Absolutely. But—frankly—such "fog of war" issues are not much different than those military commanders routinely confront in the other domains of conflict where difficult decisions frequently must be made on imperfect information.

The ability (or inability) to determine facts is not a legal issue, but as much a technical problem for the specialists to solve. So if there is a difficulty in that regard, the complaint ought to be directed at cyber scientists or even policy strategists, but not the lawyers. Sure, the law requires an ability to determine the source of an attack before launching a military response, *but so does good sense and effective military strategy.*

The same can be said for the legal requirement to assess the impact on civilians and civilian objects before launching a cyber attack. This is information that decision makers would want for political and policy reasons wholly independent of any legal requirements. As the great strategist Carl von Clausewitz observed, "War is the continuation of policy by other means."[32] Again, if the ability to

---

30. Donna Miles, *Doctrine to Establish Rules of Engagement Against Cyber Attacks*, AMERICAN FORCES PRESS SERVICE, Oct. 20, 2011, *available at* http://www.defense.gov/news/newsarticle.aspx?id=65739.

31. Gen. Charles Dunlap, *Perspectives for Cyber Strategists on Law for Cyberwar*, STRATEGIC STUDIES Q. (Spring 2011), *available at* http://www.au.af.mil/au/ssq/2011/spring/dunlap.pdf.

32. Christopher Bassword, *Clausewitz and His Works*, Clausewitz.com (updated July 8, 2008), http://www.clausewitz.com/readings/Bassford/Cworks/Works.htm.

make the calculations that political leaders and policy makers require as much as lawyers is inadequate, that is a technical, *not* legal, issue.

When—and if—the facts and circumstances are determined, weighing them is what policy makers and military commanders "do." Lawyers may help them, but ultimately it is the decision maker's call, not the lawyer's. Any reluctance of decision makers to make difficult fact determinations—if such reluctance does exist—is *not*, in any event, a deficiency of *law*, but of *leadership*.

Of course, such decisions are never exclusively about legal matters. Policy makers and commanders rightly take into account a variety of factors beyond the law. In actual practice, it appears that such considerations often are more limiting than the law.

For example, the *Washington Post* reported that U.S. cyber weapons "had been considered to disrupt Gaddafi's air defenses" early in NATO's UN-sanctioned operations aimed at protecting Libyan civilians.[33] However, the effort "was aborted," the *Post* said, "when it became clear that there was not enough time for a cyberattack to work." Conventional weapons, it was said, were "faster, and more potent," a pure *military* rationale.

None of this reflects even the slightest suggestion that "lawyers" or the law frustrated the execution of a cyber operation in Libya.

No doubt there was discussion about cyber-reporting obligations under the War Powers Resolution, but Presidents have almost never seen that as a bar to military actions, so it can hardly be said to be something unique to cyber operations or that operated to actually block a cyber attack, per se. Rather, it is but one of the many *political* considerations applicable to military actions generally, cyber or otherwise.

To be clear, the primary concern about the potential use of cyber weaponry against Libya was *not* anything generated by lawyers as

---

33. Ellen Nakashima, *U.S. Cyberweapons Had Been Considered to Disrupt Gaddafi's Air Defenses*, Wash. Post (Oct. 17, 2011), *available at* http://www.washingtonpost.com/world/national-security/us-cyber-weapons-had-been-considered-to-disrupt-gaddafis-air-defenses/2011/10/17/gIQAETpssL_story.html?wprss=rss_national-security.

Baker might put it, but rather by "administration officials and even some military officers" who, the *New York Times* says, "balked, fearing that it might set a precedent for other nations, in particular Russia or China, to carry out such offensives of their own." Along this line, the *Times* quoted James Andrew Lewis, a senior fellow at the Center for Strategic and International Studies, as opining that the United States does not want to be the "ones who break the glass on this new kind of warfare."[34]

Again, the legitimacy of these concerns aside, they illustrate—regardless—that while there may be unresolved *policy* questions inhibiting cyber operations, that is altogether different from the *legal* problems of Baker's imaginings.

The threat of cyberwar is certainly an extremely serious one, but surely not a greater peril than is nuclear war. Yet at least insofar as the U.S. military is concerned, nuclear operations *can* be made amenable to the law.[35] In other words, if our survival does not require abandoning the rule of law with respect to nuclear weapons, there is certainly no reason to do so in the cyber realm.

Does Baker nevertheless believe that the United States is so vulnerable to catastrophic cyber attack that the nation must reject any legal limits in its cyber response?

If, indeed, the United States were as vulnerable to *catastrophic* attack as Baker would have us believe, al Qaeda or some extremist group certainly would have launched one by now. In point of fact, although cyber crime may be extensive, militarily significant cyber attacks apparently are not so easy to conduct as Baker seems to think. In reporting the rejection of cyber weaponry as a means of dismantling Libyan air defenses, *The New York Times* noted that:

While popular fiction and films depict cyberattacks as easy to mount—only a few computer keystrokes needed—in reality it takes

---

34. Eric Schmitt & Thom Shanker, *U.S. Debated Cyberwarfare in Attack Plan on Libya*, N.Y. Times (Oct. 17, 2011), *available at* http://www.nytimes.com/2011/10/18/world/africa/cyber-warfare-against-libya-was-debated-by-us.html?_r=2.

35. Col. Charles Dunlap, *Taming Shiva: Applying International Law to Nuclear Operations*, Air Force L. Rev. (Vol. 42, 1997), *available at* http://www.afjag.af.mil/shared/media/document/AFD-081204-037.pdf.

significant digital snooping to identify potential entry points and susceptible nodes in a linked network of communications systems, radars and missiles like that operated by the Libyan government, and then to write and insert the proper poisonous codes.[36]

Obviously, if cyber weaponry is technically difficult for the world's foremost military to use even against a third-world power such as Libya, one may reasonably infer that it is markedly more difficult to use against a sophisticated first-world power, even for a peer or near peer of that power.

Rejection of legal limits carries other, real-world consequences that are not in the United States' cyber interests. An effective response to cyber threats is not an autarchic enterprise; it requires the cooperation of international allies. Baker's "damn the law and lawyers" approach would cripple our relations with the law-abiding nations whose cooperation we must have to address cyber threats.

We need to keep in mind that the vast majority of adverse cyber incidents are criminal matters, and the resolution of them frequently necessitates the involvement of foreign police and judicial authorities who, by definition, require partners who are themselves committed to faithfulness to the rule of law.

The importance of legal legitimacy cannot be overstated. As outlined above, few in uniform who have experienced the vicissitudes of war since 9/11 would underestimate the deleterious impact on coalition support that the mere *perception* of American lawlessness can have.

In any event, the American people insist upon legality. Michael Reisman and Chris T. Antoniou noted in 1994 that the public support that democracies need to wage war "can erode or even reverse itself rapidly, no matter how worthy the political objective, if people *believe* that the war is being conducted in an unfair, inhumane, or iniquitous way."

---

36. Eric Schmitt & Thom Shanker, *U.S. Debated Cyberwarfare in Attack Plan on Libya*, N.Y. TIMES (Oct. 17, 2011), *available at* http://www.nytimes.com/2011/10/18/world/africa/cyber-warfare-against-libya-was-debated-by-us.html?_r=2.

In truth, as important as the moral perspective may be, the *practical* advantages of adherence to the rule of law have a power all their own—as history plainly shows.

Nazi Germany's and Imperial Japan's gruesome violations of the law of war, for example, hardly proved advantageous to them. More recently, Saddam Hussein, who embraced war without "limits," was pulled from a subterranean spider hole, dirty, defeated, and soon-to-be-dead.[37] Muammar Gaddafi's illicit threats to wage war upon his own civilian population in the spring of 2011 brought the military power of the international community down upon him to the point where he ended his days groveling in a sewer pipe.[38]

Military leaders know that adherence to the law is a pragmatic essential to prevailing in 21st-century conflicts. It might be attractive to some to capitalize on the unpopularity of lawyers to demonize them, and even the law itself, but military commanders understand that war today has changed. They know that law has permeated war much as it has every other human activity, and they realize the perils of ignoring its power and influence. Whether anyone likes it or not, war has become, as General James Jones, then the commander of NATO forces, observed in 2003, "very legalist and very complex."[39]

And lawyers? "Now," Jones said, "you need a lawyer or dozen." To which one might today add "if you want to win."

Cyberwar: Reply to General Dunlap

Stewart Baker

General Dunlap's essay has two broad themes.

---

37. Frances Romero, *Saddam Hussein's Spider Hole*, Time, May 4, 2011, *available at* http://www.time.com/time/specials/packages/article/0,28804,2069355_2069356_2069377,00.html.

38. "Security Council Approves 'No-Fly Zone' over Libya, Authorizing 'All Necessary Measures' to Protect Civilians, by Vote of 10 in Favour with 5 Abstentions," U.N. Security Council (March 17, 2011), *available at* http://www.un.org/News/Press/docs/2011/sc10200.doc.htm.

39. Gen. James L. Jones, *A Marine's Toughest Mission*, Parade Magazine (accessed Jan. 26, 2012), *available at* http://www.parade.com/articles/editions/2003/edition_01-19-2003/General_Jones.

First, he argues that a "lawyers-first" approach to cyberwar won't really handicap our military. I agree with some of what he says. The United States does have formidable offensive cyber weapons, and we should benefit from ambiguity about how we will respond if attacked.

But on other fronts, he's almost certainly wrong, as when he suggests that America's military strength will deter cyber attacks, or that cyber weapons aren't that big a deal, since we didn't use them against Libya. We didn't use such weapons against Libya in part because we didn't have enough time to map Libya's infrastructure and in part because we didn't want to encourage the routine military use of cyber weapons. But neither of those considerations is likely to deter our adversaries, who have been mapping the U.S. infrastructure for years and who will no doubt launch their attacks anonymously and deniably. What good will all our offensive weapons do us then? At worst, we'll be reduced to raging helplessly, unsure of whom to attack. At best, we'll be started on a path taken by the bombers in World War II—a cycle of escalating attacks and counterattacks that will quickly destroy the notion that there are legal limits on civilian targeting.

General Dunlap's second theme is plainly heartfelt but equally mistaken. To him, taking lawyers out of cyberwar strategy will lead to "lawless war," and he pulls out all the stops to condemn it, invoking Abu Ghraib, Adolf Hitler, Imperial Japan, and, um, Alberto Gonzales.

If you're wondering how the former Attorney General got on that list, I suspect it's because General Dunlap is still fighting the last war. The last turf war, to be precise. The years after 9/11 saw bitter conflict between military judge advocate generals and civilian leaders like Alberto Gonzales. They fought over military tribunals, Guantánamo, and interrogation.

The military lawyers mostly won. But the cost of that victory was high. It did surprising damage to civilian control of the military (it's hard, for example, to read General Dunlap's essay without getting the impression that "civilian lawyer" is some new kind of epithet). And it led military and national security lawyers to draw the wrong lessons from the post–9/11 wars. In the future, they concluded, no war should be planned or fought without a lawyer at every commander's elbow.

Really? Let's assume, despite substantial contrary evidence, that when we fight in places such as Libya or Iraq or Afghanistan, we can deprive our adversaries of propaganda victories so long as our military does nothing without a lawyer's approval. Even if that's true, why would we expect the same approach to work for a war in cyberspace?

At its worst, cyberwar could reduce large parts of the United States to the condition of post-Katrina New Orleans, maybe for weeks or months. Responding to propaganda attacks isn't likely to be high on our to-do list. Indeed, we have not planned for a war with such dire domestic consequences since the 1950s, when atomic weapons, long-range bombers, and intercontinental missiles transformed our military strategy.

For the first time since the 1950s, we must recognize something no lawyer likes to admit:

> that law has only a limited role in fighting wars where national survival is on the line. That's the lesson from FDR's failed effort to outlaw air attacks on civilian targets. Or, really, from dozens of other episodes. In 1936, submarine stealth attacks on merchant ships were violations of the law of war—a rule that was swept away in the first hours of World War II. In World War I, much the same fate befell a solemn international condemnation of poison gas weapons.

We knew this lesson once. At Nuremberg, we declined to charge German submarine and air commanders who had attacked civilian targets, recognizing that the illusory law of the 1930s had been nullified on the battlefield.

Our first nuclear strategy was grounded in that hard-won experience. We didn't give the JAG corps a veto over nuclear weapons or strategy. We didn't station lawyers in the silos to decide when the missiles could be launched. And I am confident there's no legal memo in the Pentagon's files from that era saying, "We think it's perfectly proportional and lawful to respond to a conventional Soviet tank advance in Germany by launching a massive nuclear attack over the airspace of unwilling neutral nations, incinerating millions of Russian civilians, and covering the world in radioactive fallout."

We came up with that admittedly ugly nuclear strategy not because the lawyers liked it—how could they?—but because it was the only way to save Europe from a Soviet invasion. Oh, and because it worked.

That's the kind of realism the government will need as it plans for the challenge of cyberwar.

It is not yet a kind of realism that is easy to find, even in the Pentagon.

Cyberwar: Reply to Stewart Baker

Charles J. Dunlap Jr.

My friend Stewart Baker's vociferous response to my critique of his polemic is an exquisite example of something cleverly strategized to play well in the gilded salons of certain Washington *glitterati*, but not so much in the stark command posts of those tasked with real responsibility for America's cyber security.

By bashing military lawyers (and others) who have stood up—and who will continue to stand up—for the rule of law, Baker seems to think he will amass converts to his strange belief that lawlessness wins wars.

Don't drink the Kool-Aid. People with concrete experience in actual conflicts know well what happens these days to those who think that the law doesn't matter. They end up not just as battlefield losers, but also with a hangman's noose around their neck (Saddam) or bullets in their forehead (Bin Laden) or cowering in a sewer pipe (Gaddafi).

Baker clings to his dogma that the only way to deter cyber attacks is to threaten innocent people with some kind of cyber damnation. Moral considerations aside, the 21st century is replete with examples that prove that too many of our most dangerous adversaries are rather indifferent to the fate of civilians, including their own people.

Does anyone think that those disposed to set off bombs in markets crowded with children would be deterred by a threat to cybernetically shut down hospital incubators somewhere? Cruelly enough, such adversaries just don't care that much about dead babies. It really is that simple.

What *military* experience shows, however, is that what *might* actually work is to hold at risk the perpetrators themselves and, often, the means by which they execute their attacks. That Baker so rejects this proven—and fully lawful—*military* approach is genuinely puzzling.

Perhaps it is helpful if we separate Baker's obsessive abhorrence of the legal profession generally from hostility toward the law itself. My argument is this: Whatever one might think of lawyers—military or civilian—from a *military* perspective adherence to the law itself is not just intrinsic to an honorable and decent people; it is also a practical, hard-nosed *necessity* for success in contemporary military operations.

Why? Again, the answer is uncomplicated (and one that Baker was wholly unable to counter in his response): Among other things, a damn-the-law strategy would deprive the United States of the international cooperation that countering a cyber threat (especially) absolutely requires. Few serious observers of the post-9/11 world dispute this axiom.

Another problem with his theory arises from Baker's apparent misreading of the ethic of those in uniform who would be called on to execute the kind of cyber attack against noncombatants that so enthralls him.

Members of the armed forces take their oath to support and defend the Constitution very seriously. They do not enlist to conduct

plainly unlawful operations against the helpless, however ideologi-
cally popular doing so might be in various quarters.

Consequently, it is absurd to think that those in the military would,
in any event, knowingly embrace the type of illicit stratagem he
proposes. Wearing the uniform does not transform Americans into
automatons indifferent to the rule of law . . . as some evidently
assume is the case.

Furthermore, it is obvious that the American people would back
their military's ethic, and not Baker's philosophy. Professors Michael
Riesman and Chris T. Antoniou point out in their 1994 book, *The
Laws of War*:

> In modern popular democracies, even a limited armed con-
> flict requires a substantial base of public support. That sup-
> port can erode or even reverse itself rapidly, no matter how
> worthy the political objective, if people believe that the war
> is being conducted in an unfair, inhumane, or iniquitous way.

Baker's theory is further unraveled by the military reality that
those adversaries who are deterrable at all—typically nation-states—
are just as (or *more*) likely to be deterred by the unambiguous cer-
tainties of the United States' amply demonstrated non-cyber military
might as by anything that might be done in the cyber realm, which is
still littered with uncertainties and ambiguities.

In trying to counter this indisputable truth, Baker's logic crumbles.
For example, he alleges that adversaries can attack us anonymously
and leave us "reduced to raging helplessly, unsure of whom to attack."

If adversaries really can do that, how does threatening innocents
with unrestricted cyberwar change the calculation? Indeed, if we can
be rendered truly helpless by a cyber first strike by an enemy confident in his ability to remain anonymous, how would threats—of *any*
kind—deter him?

Oddly, Baker does not stick with his reduced-to-helplessness es-
timate. He later concedes that "[a]t its worst, cyberwar could reduce

large parts of the United States to the condition of post-Katrina New Orleans."

The fact is that even if Baker's very doubtful worst-case scenario came to pass, and even if it also happened that "large parts" of the American military were similarly humbled, the U.S. arsenal is so vast and formidable that the remaining portion would still comprise a terrifying force well able to devastate any nation on the planet.

Military planners around the globe keenly appreciate this truth, even if others do not.

Finally, Baker seeks to analogize his "lawless" concept of cyber deterrence with nuclear deterrence because, it seems, he assumes that our nuclear forces operate outside the law. That is incorrect, and I invite his attention to my 1997 article, "Taming Shiva: Applying International Law to Nuclear Operations," *where I explain the legal review process in which I personally participated.*[40]

No doubt some armchair "Rambos" will find Baker's forget-the-law theory alluring, but from a military perspective, there is no question that the adherence to the rule of law is not just the right thing to do, it is also the smart, pragmatic, and *war-winning* formula.

---

40. Charles Dunlap, *Taming Shiva: Applying International Law to Nuclear Operations*, AIR FORCE L. REV. (vol. 42, 1997), *available at* http://www.afjag.af.mil/shared/media/document/AFD-081204-037.pdf.

Chapter Ten
Detention Policies

Stephen I. Vladeck

Greg Jacob

Introduction

The restrictions on the transfer of detainees to the custody or effective control of foreign countries interfere with the authority of the executive branch to make important and consequential foreign policy and national security determinations regarding whether and under what circumstances such transfers should occur in the context of an ongoing armed conflict . . . Nevertheless, my Administration will work with the Congress to seek repeal of these restrictions, will seek to mitigate their effects, and will oppose any attempt to extend or expand them in the future.[1]

—President Barack H. Obama,
Signing Statement on H.R. 6523 (the National Defense
Authorization Act for 2011), January 7, 2011

The transfer of [detainees to the United States] directly contradicts Congressional intent and the will of the American people . . . Congress has spoken clearly multiple times—

---

1. President Barack H. Obama, Statement by the President on H.R. 6523, Jan. 7, 2011 (http://www.whitehouse.gov/the-press-office/2011/01/07/statement-president-hr-6523).

including explicitly in pending legislation—of the perils of bringing terrorists onto U.S. soil. It is unacceptable that the administration notified Congress only after it unilaterally transferred [Ahmed Warsame] to New York City despite multiple requests for consultation.[2]

—Statement of Representative Howard "Buck" McKeon
(R-CA), July 5, 2011

The Future of Military Detention: Guantánamo and Beyond

Stephen I. Vladeck

Writing for a divided panel of the D.C. Circuit in *Latif v. Obama*,[3] Judge Janice Rogers Brown openly criticized the U.S. Supreme Court's decision in *Boumediene v. Bush,* which held that the Constitution's Suspension Clause entitles non-citizens detained at Guantánamo to meaningful judicial review of the legality of their ongoing confinement.[4] In her words,

> *Boumediene* fundamentally altered the calculus of war, guaranteeing that the benefit of intelligence that might be gained— even from high-value detainees—is outweighed by the systemic cost of defending detention decisions. While the court in *Boumediene* expressed sensitivity to such concerns, it did not find them "dispositive." *Boumediene*'s logic is compelling: take no prisoners. Point taken.[5]

Judge Brown's rhetoric provides a useful lens for thinking about the future of U.S. detention policy, for it can fairly be seen as suggest-

---

2. Charlie Savage & Eric Schmitt, *U.S. to Prosecute a Somali Suspect in Civilian Court*, N.Y. Times, July 5, 2011 (http://www.nytimes.com/2011/07/06/world/africa/06detain.html).

3. No. 10-5319, 2011 WL 5431524 (D.C. Cir. Oct. 14, 2011).

4. 553 U.S. 723 (2008).

5. *Latif*, 2011 WL 5431524, at *15 (citation omitted).

ing that the Supreme Court's various interventions into detainee policy in the War on Terrorism have been directly responsible for the "shrinking category of cases" arising out of Guantánamo and the related reality that "[t]he ranks of Guantanamo detainees will not be replenished."[6] Put another way, faced with the specter of judicial review, *Latif* suggests that the Bush and Obama administrations were compelled to resort to other measures for handling terrorism suspects, whether detention at other overseas locations (to which the Suspension Clause might not run); indictment and trial by civilian U.S. courts; or more lethal forms of incapacitation—including targeted killings. Indeed, if Judge Brown is right, then the result would be profoundly unsettling: The true lesson of the past decade with regard to military detention is that judicial review is ultimately self-defeating, provoking responses by the political branches that largely eliminate the need for (or availability of) judicial review in future cases.

The short chapter that follows aims to take Judge Brown's suggestion seriously. As I explain, although Judge Brown is clearly correct that judicial review has impacted the *size* of the detainee populations within the territorial United States and at Guantánamo, it does not even remotely follow that the jurisprudence of the past decade has precipitated a shift away from detention and toward targeted killings. To the contrary, the jurisprudence of Judge Brown's own court has simultaneously (1) left the government with far *greater* detention authority than might otherwise be apparent where noncitizens outside the United States are concerned; and (2) for better or worse, added a semblance of legitimacy to a regime that had previously and repeatedly been decried as lawless. And in cases where judicial review prompted the government to release those against whom it had insufficient evidence, the effects of such review can only be seen as salutary. Thus, at the end of a decade where not a single U.S. military detainee was freed by order of a federal judge, it is more than a little ironic for Judge Brown to identify "take no prisoners" as *Boumediene*'s true legacy.

---

6. *Id.*

I

The role of judicial review in the three post–9/11 military detention cases in which the detainees were held within the territorial United States is impossible to overstate. Despite the Bush administration's initial position that the detention of "enemy combatants" posed a non-justiciable political question,[7] the federal courts (and the Supreme Court, in particular) were emphatic in suggesting that such detentions *were* subject to judicial review, even as they divided over the merits in each of the three cases.

Thus, in the case of Yasser Esam Hamdi, the federal government argued to the Supreme Court that "some evidence" was sufficient to justify the long-term detention of U.S. citizens captured on the battlefield. Although the Court agreed that the government had the authority to detain individuals *like* Hamdi,[8] it disagreed as to the evidentiary burden, with a 6-1 majority concluding that a more rigorous evidentiary burden was necessary.[9] Rather than attempting to provide such evidence on remand, the government quickly entered into an agreement with Hamdi wherein he agreed to relinquish his citizenship in exchange for his release and transfer to Saudi Arabia.[10]

In the case of Jose Padilla, although the Supreme Court initially threw out Padilla's habeas petition in 2004 on the ground that he had filed in the wrong district court,[11] the opinions in the contemporaneous *Padilla* and *Hamdi* decisions left the distinct impression that, on the merits, five Justices would have rejected the argument that the 2001 Authorization for the Use of Military Force[12] authorized the detention of U.S. citizens arrested within the territorial United States.[13]

---

7. *See, e.g.*, Hamdi v. Rumsfeld, 296 F.3d 278, 283 (4th Cir. 2002).

8. Hamdi v. Rumsfeld, 542 U.S. 507, 516–24 (2004) (plurality opinion).

9. *See id.* at 524–39; *id.* at 553–54 (Souter, J., concurring in part, dissenting in part, and concurring in the judgment).

10. *See* Jerry Markon, *Hamdi Returned to Saudi Arabia*, WASH. POST, Oct. 14, 2004, at A2.

11. *See* Rumsfeld v. Padilla, 542 U.S. 426 (2004).

12. Pub. L. No. 107-40, 115 Stat. 224 (codified at 50 U.S.C. § 1541 note).

13. *See, e.g.*, Stephen I. Vladeck, *The Long War, the Federal Courts, and the Necessity/Legality Paradox*, 43 U. RICH. L. REV. 893, 910 & n.105 (2009).

Padilla refiled in the proper venue,[14] only to have the government moot the case on the eve of the Supreme Court's review by indicting him on criminal charges and transferring him to civilian custody. As Fourth Circuit Judge J. Michael Luttig observed, the timing of the government's conduct gave rise "to at least an appearance that the purpose of these actions may be to avoid consideration of our decision [upholding Padilla's detention] by the Supreme Court."[15] Nevertheless, and over three dissents, the Court denied certiorari.[16]

That pattern repeated itself in the case of Ali al-Marri (the one non-citizen subjected to military detention within the territorial United States), with the Obama administration mooting the merits of his detention after the Supreme Court granted certiorari[17] by indicting him on criminal charges and transferring him to civilian custody.[18] Thus, in all three cases, the specter of future judicial review—in the district court in *Hamdi* and in the Supreme Court in *Padilla* and *al-Marri*—directly led to a change in policy, and there have been no additional stateside military detention cases since.

II

At least based on the public record, one can only make an inferential case that this pattern has repeated with regard to Guantánamo, but the circumstantial evidence is fairly compelling. Although 779 non-citizens were at one time detained as "enemy combatants" at Guantánamo, the detainee population dropped from 597 at the time of the Supreme Court's *Rasul* decision in 2004 to 269 at the time *Boumediene* was decided, and from there to the 171 men detained there today.[19] And although none of the 600 detainees who have been

---

14. *See* Padilla v. Hanft, 389 F. Supp. 2d 678 (D.S.C.), *rev'd*, 423 F.3d 386 (4th Cir. 2005).

15. Padilla v. Hanft, 432 F.3d 582, 585 (4th Cir. 2005).

16. *See* Padilla v. Hanft, 547 U.S. 1062 (2006) (mem.).

17. *See* al-Marri v. Pucciarelli, 534 F.3d 213 (4th Cir.) (en banc), *cert. granted*, 555 U.S. 1066 (2008).

18. *See* al-Marri v. Spagone, 555 U.S. 1220 (2009) (mem.).

19. For the data, see *A History of the Detainee Population*, N.Y. Times, http://projects.nytimes.com/guantanamo.

released from Guantánamo were directly freed by a judicial order, it stands to reason that the sharp uptick in the rate of transfers *out* of Guantánamo (along with the virtual cessation of transfers in) after June of 2004 was a direct reaction to, and result of, the Court's decision in *Rasul v. Bush,* which held that the federal habeas statute extended to Guantánamo.[20] Moreover, in the four years since *Boumediene,* there have been at least 11 distinct district court decisions granting habeas relief that the government declined to appeal on the merits.[21] Not all the detainees at issue in those cases have been released,[22] but those that were certainly weren't *hurt* by the judicial proceedings on their behalf.

Inasmuch as the detainee litigation appears to have exerted hydraulic pressure on the Executive Branch to reduce the detainee population at Guantánamo, it has arguably also invested the detentions in the cases that remain with at least a modicum of legitimacy—at least for those detainees who have not been cleared for release. After all, the government now is able to argue that the detainees still at Guantánamo have received the exact judicial review called for by the Constitution; the fact that the courts have *denied* relief in many of those cases only underscores the validity of that aspect of the U.S. detention regime in the short term (and perhaps in the long term, as well).

III

Far less data exists to evaluate the relationship between judicial review and the number of detainees held by the United States in Afghanistan.[23] Here, though, the data is less important than the case law. Notwithstanding *Boumediene,* the D.C. Circuit held in *al-Maqaleh v. Gates* that non-citizens detained in Afghanistan, even if they are not citizens of *or* arrested in Afghanistan, are not entitled to

---

20. 542 U.S. 466 (2004).

21. *See* Stephen I. Vladeck, *The D.C. Circuit After* Boumediene, 41 SETON HALL L. REV. 1451, 1474 & n.127 (2011).

22. *See id.* at 1476–88.

23. *See* Jeff A. Bovarnick, *Detainee Review Boards in Afghanistan: From Strategic Liability to Legitimacy,* ARMY LAW., June 2010, at 9.

pursue habeas relief in the U.S. federal courts.[24] In so holding, the court of appeals specifically rejected the detainees' argument that judicial review must be available lest the government deliberately choose to send new detainees to Afghanistan in order to escape judicial oversight:

> [T]he notion that the United States deliberately confined the detainees in the theater of war rather than at, for example, Guantanamo, is not only unsupported by the evidence, it is not supported by reason. To have made such a deliberate decision to "turn off the Constitution" would have required the military commanders or other Executive officials making the situs determination to . . . predict the *Boumediene* decision long before it came down.[25]

Because *Maqaleh* means that judicial review will not extend to Afghanistan absent a showing of deliberate manipulation on the government's part (and perhaps not even then), the conclusion appears manifest that *Boumediene*'s holding is limited to Guantánamo, and that the government, in fact, does not face the prospect of judicial review in future cases involving the detention of non-citizens elsewhere outside the territorial United States. As such, Judge Brown's suggestion in *Latif* that *Boumediene* has chilled (and will chill) future military detentions of terrorism suspects necessarily fails to persuade. At least for non-citizens picked up outside the territorial United States, *Maqaleh* preserves substantial flexibility on the government's part, and leaves judicial review as an unlikely proposition, at best.

*   *   *

But there's another aspect to the jurisprudence of the past decade that also poses a stark contrast with Judge Brown's reasoning: Thanks to the work of Judge Brown and her colleagues on the D.C. Circuit, even in cases in which judicial review *does* apply, the relevant sub-

---

24. 605 F.3d 84 (D.C. Cir. 2010).
25. *Id.* at 99.

stantive and procedural standards governing such review leave the government with sweeping authority.[26] With regard to non-citizens outside the territorial United States, current case law requires the government to show merely by a preponderance of the evidence (i.e., that it is more likely than not) that the detainee was "part of" or "substantially supported" al Qaeda. And thanks to *Latif* (the very decision in which Judge Brown objected to *Boumediene*), intelligence reports are treated with a presumption of regularity—making it incredibly difficult as a practical matter for detainees to overcome the government's evidence.[27] In point of fact, there has not been a single case to date in which the D.C. Circuit either affirmed a district court's grant of habeas relief or reversed the denial thereof. Given the government's successful track record before Judge Brown and her colleagues, it's that much harder to understand her claim that "the systemic cost of defending detention decisions" has dissuaded the government from doing so. If the litigation of the past few years has suggested anything with regard to the future of U.S. detainee policy, it is that the cost to the government of defending detention decisions in the D.C. Circuit is not particularly high, especially compared to the benefit that such review has provided.

The Right Balance: Judicial Humility in Times of War

Greg Jacob

In arguing that the Supreme Court's decision in *Boumediene v. Bush*[28] has not "precipitated a shift away from detention and toward targeted killings," Professor Vladeck knocks down an easily-dispensed-with straw man, but fails to tackle the more interesting question of whether the D.C. Circuit's post-*Boumediene* jurisprudence has struck the right balance in establishing parameters for judicial review of

---

26. *See* Vladeck, *supra* note 21.
27. *See* Latif v. Obama, No. 10-5319, 2011 WL 5431524, at *20–41 (Tatel, J., dissenting).
28. 553 U.S. 723, 128 S. Ct. 2229 (2008).

executive branch decisions concerning the detention of captured enemy combatants. This short article suggests that, in at least two significant respects, it has. First, the D.C. Circuit's decision in *al-Maqaleh v. Gates*[29] avoids disruptive litigation over the military detention of most aliens who are (1) captured abroad, (2) designated by the military as enemy combatants, and (3) held in a theater of active military operations. Second, the evidentiary burdens and presumptions applied by the D.C. Circuit in reviewing the habeas petitions of Guantánamo detainees have, by and large, struck an appropriate balance between the "practical considerations and exigent circumstances"[30] of needing to avoid "judicial interference with the military's efforts to contain 'enemy combatants [and] guerilla fighters,'"[31] on the one hand, and the need to "protect against the arbitrary [and unlawful] exercise of governmental power" on the other.[32]

Boumediene and Its Progeny: New Law for a New Kind of Conflict

Captured enemy combatants, whether lawful or unlawful, are not detained for the purpose of punishment, but rather to prevent them from rejoining enemy forces and engaging in further hostilities. Such detention authority is no less necessary in a guerrilla war against covert terrorist elements than it is in large-scale conventional conflicts. If our military forces are competent—and they certainly are—circumstances will arise in which hard-pressed enemy forces will elect to lay down their arms and voluntarily surrender. And if we as a people are both moral and merciful—and we strive to be——rather than kill the surrendered enemy, we will instead offer to detain them.

But what then? The Supreme Court's decisions in *Ex Parte Quirin*[33] and *Johnson v. Eisentrager*,[34] together with long-standing historical

---

29. 605 F.3d 84 (D.C. Cir. 2010).
30. *Boumediene*, 128 S. Ct. at 2275.
31. *Id.* at 2261, *quoting* Johnson v. Eisentrager, 339 U.S. at 763, 784 (1950).
32. *Id.* at 2275.
33. 317 U.S. 1 (1942).
34. 339 U.S. at 763 (1950).

practice, establish the government's authority to hold captured enemy combatants until the end of an armed conflict to prevent them from rejoining the fray and attempting to kill our forces. But does this well-established rule apply without limitation in an armed conflict that had no natural end? On the one hand, releasing an avowed enemy of the United States, whose hatred of our country can only have been inflamed by years of detention, and without any firm assurance that he or she will not seek to engage in future hostilities against us, seems the sheerest of folly. On the other hand, if it will never be assuredly safe to release such individuals, can we really preventively detain them indefinitely, possibly for decades, and perhaps even until the end of their natural lives? Historical and legal precedents have almost all described the government's authority to detain enemy combatants in absolute terms, but those precedents have seemed to assume that the underlying conflicts would eventually end, and that the government's detention authority would thus come to a natural close.

In the early War on Terror cases, the government subscribed to the absolute theory of detention authority that flowed from these precedents. And in the Supreme Court's first examination of a War on Terror case, it agreed, holding that "universal agreement and practice" support the military's authority to capture and detain individuals who are "part of or supporting forces hostile to the United States . . . and engaged in armed conflict against the United States."[35] The Supreme Court expressly noted that the purpose of such detention is to prevent enemy combatants from "returning to the field of battle and taking up arms once again," stating that combatants can accordingly be held "for the duration of the relevant conflict."[36]

More than a decade into the War on Terror, no federal court has seriously called into question the government's potentially unending authority to detain captured War on Terror combatants until the con-

---

35. Hamdi v. Rumsfeld, 542 U.S. 507, 516 (2004). The Court expressly held that this detention authority extends even to U.S. citizens.

36. *Id*. at 518.

flict "ends."[37] Whether there are or should be any temporal limitations to that authority is a question that future judges and political leaders may well address.[38] *Boumediene,* however, demonstrates the judiciary's concern that as the War on Terror drags on, and with it the length of ongoing detentions (at the time of the *Boumediene* decision, some of the detainees had been held for more than six years), we need to at least be increasingly sure that the individuals we are detaining are, in fact, enemy combatants. *Boumediene* expressly declined to state how greater certainty concerning the validity of military detentions should be achieved, noting that "our opinion does not address the content of the law that governs [enemy combatant] detention" and directing the lower courts to establish a framework capable of reconciling "[l]iberty and security . . . within the framework of the law."[39] This is what the D.C. Circuit has attempted to do.

Avoiding Judicial Interference in Theaters of Active Hostilities

Probably the most important War on Terror decision handed down by the D.C. Circuit since *Boumediene* was decided is *Maqaleh,* in which the Court declined to extend the writ of habeas corpus to aliens captured abroad, designated enemy combatants, and held at Bagram Air Force Base in Afghanistan. From the military's perspective, the nightmare scenario has always been the prospect that the judiciary would assert the right to engage in a searching inquiry into the basis for every capture and detention of an alien abroad, even

---

37. Because Congress passed the Authorization for Use of Military Force almost immediately after the September 11 attacks, P.L. 107-40, 115 Stat. 224 (2001), War on Terror decisions have not needed to address the extent to which the President may have inherent authority as Commander in Chief, even absent congressional authorization, to detain captured enemy combatants.

38. "Because our Nation's past military conflicts have been of limited duration, it has been possible to leave the outer boundaries of war powers undefined. If, as some fear, terrorism continues to pose dangerous threats to us for years to come, the Court might not have this luxury." *Boumediene,* 128 S. Ct. at 2277.

39. *Id.*

while active combat operations are ongoing. In World War II, such a rule could have required the government to litigate hundreds of thousands of habeas claims, costing the government significant expense and causing substantial disruption to military operations. *Maqaleh* puts such fears to rest.

In declining to exercise habeas jurisdiction over Bagram, the *Maqaleh* court did not apply a bright-line territorial sovereignty test, but rather engaged in a multifactor analysis drawn from *Boumediene* that examines (1) the citizenship and status of the detainee and the adequacy of the process through which the status determination was made, (2) the nature of the site of apprehension and the site of detention, and (3) the practical obstacles inherent in resolving the prisoner's entitlement to the writ. The essential holding of the case seems to be that where the government apprehends an alien abroad, and then detains that alien at a location not within the de jure or de facto sovereignty of the United States and within an active theater of war, the writ of habeas corpus does not apply. There may be exceptions to this rule where the government has not engaged in any formal process for determining whether detained individuals are legitimately classified as enemy combatants,[40] or where the government has deliberately transferred prisoners to an active theater of war for the purpose of avoiding habeas jurisdiction.[41] Otherwise, however, *Maqaleh* requires the judiciary to exercise some humility and defer to most military detention decisions in active theaters of war.

Does the *Maqaleh* rule create the possibility, and perhaps even the likelihood, of erroneous detentions? Certainly. Mankind has not yet devised a perfect system for correcting such errors. But the decision is founded on a principle long recognized by the courts: That absent extraordinary circumstances, the cost to security of judicial interference in active overseas military operations outweighs the liberty cost of potentially erroneous detentions pursuant to those operations. Thus, five years after World War II formally ended, the Supreme Court declined to extend the writ of habeas corpus to prison-

---

40. *Mawaleh*, 605 F.3d at 96.
41. *Id.* at 98–99.

ers held in Germany, explaining that "[s]uch trials would hamper the war effort and bring aid and comfort to the enemy. . . . It would be difficult to devise more effective fettering of a field commander than to allow the very enemies he is ordered to reduce to submission to call him to account in his own civil courts and divert his efforts and attention from the military offensive abroad to the legal defensive at home."[42] Through *Maqaleh,* these legitimate concerns continue to govern the enemy combatant jurisprudence of today.

Evidentiary Burdens and Presumptions

The law developed by the D.C. Circuit for reviewing enemy combatant habeas petitions is perhaps more important for its symbolic than its practical effect, as the government has stopped transferring detainees to Guantánamo Bay, and the number of individuals to which the body of law applies is thus small and shrinking. Nevertheless, we are a nation that prizes liberty, fairness, and the rule of law, and the *Boumediene* court may well have been correct that outside the theater of war, the liberty cost of potentially erroneous detentions must under some circumstances outweigh the security costs attendant to habeas review.

The key question in a habeas inquiry concerning a captured enemy combatant is not whether the individual committed a crime, but rather whether the government properly designated the individual an enemy combatant. This inquiry requires review of highly classified intelligence reports, which often contain hearsay statements gathered from a variety of intelligence sources. The D.C. Circuit has established that hearsay evidence is admissible in detainee review cases,[43] that the recording of hearsay statements by government agents is entitled to a presumption of regularity (but not to a presumption that the recorded hearsay statements are actually true),[44] that the various items of evidence used by the government to support a detention

---

42. *Eisentrager,* 339 U.S. at 779.
43. Al-Bihani v. Obama, 590 F.3d 866 (D.C. Cir. 2010).
44. *Latif,* 2011 WL 5431524.

must be viewed by a reviewing court as a whole, rather than in isolation,[45] and that a governmental showing by a preponderance of the evidence is sufficient to support a detention.[46] These standards, which have for the most part gained the support of judges across the D.C. Circuit's ideological spectrum, are both flexible and fair, ensuring that detainees are not held at the whim of the executive and with no supporting evidence, while recognizing that judicial review of military detentions requires some reasonable alterations to the habeas standards to which we are more accustomed.

Conclusion

Professor Vladeck misreads the import of Judge Brown's dicta in *Latif*. *Boumediene*'s message to the government, understood through the lens of the D.C. Circuit's post-*Boumediene* jurisprudence, is not "take no prisoners." Rather, it is "don't transfer prisoners to Guantánamo Bay or the territorial United States"—or, to use Judge Brown's own words, "[t]he ranks of Guantánamo detainees will not be replenished."[47]

The government created the detention facility at Guantánamo Bay so that it would have a secure location, not easily susceptible to prison breaks or the vagaries of war, at which high-value detainees could be securely held and mined for intelligence. Now that the Supreme Court has determined that keeping captured enemy combatants at Guantánamo will subject the government to the vagaries of habeas litigation, the usefulness of the facility has been substantially diminished. That does not mean, however, that the military has been incentivized to switch from capturing enemy combatants to killing them. Rather, it means that in the future, the military will simply keep detainees where it captures them, preferring the risk of prison breaks and enemy attacks to the certain cost and disruptions to intelligence gathering that are inevitably caused by repeatedly being dragged into court.

---

45. Salahi v. Obama, 625 F.3d 745 (D.C. Cir. 2010).
46. *Al-Bihani*, 590 F.3d at 877–78.
47. *Latif*, 2011 WL 5431524 at *15.

Detention Policies: Reply to Greg Jacob

Stephen I. Vladeck

Irony pervades Greg Jacob's hortatory defense of the current state of the D.C. Circuit's jurisprudence regarding U.S. detainee policy. On the one hand, Jacob sings the praises of the Court of Appeals for adopting standards that are "flexible and fair" in the Guantánamo cases,[48] and for "ensuring that detainees are not held at the whim of the executive and with no supporting evidence, while recognizing that judicial review of military detentions requires some reasonable alterations to the habeas standards to which we are more accustomed."[49] Never mind that the D.C. Circuit has yet to rule on the merits in favor of a *single* detainee (and has repeatedly reversed grants of habeas relief by the district court), or that its jurisprudence has in various places manifested thinly veiled—if not downright overt—hostility to the Supreme Court's decision in *Boumediene*.[50] From Jacob's perspective, one can look to the work of the D.C. Circuit with respect to Guantánamo as striking the "appropriate balance"[51] between the government's compelling interests and the rights (such as they are) of the detainees—and more generally as a model for how courts should approach "the vagaries of habeas litigation."[52]

And yet, at the same time, Jacob also praises the D.C. Circuit for virtually foreclosing judicial review of the detention of non-citizens anywhere else in the world in *al-Maqaleh* v. *Gates*,[53] suggesting that, "absent extraordinary circumstances, the cost to security of judicial interference in active overseas military operations outweighs the liberty cost of potentially erroneous detentions pursuant to those operations."[54] Jacob offers no evidence of the "cost to security of judicial

---

48. Jacob Response at 4.

49. *Id.*

50. *See* Vladeck Opening at 4; Stephen I. Vladeck, *The D.C. Circuit After Boumediene*, 41 SETON HALL L. REV. 1451 (2011).

51. Jacob Response at 1.

52. *Id.* at 4.

53. 605 F.3d 84 (D.C. Cir. 2010).

54. Jacob Response at 3.

interference in active overseas military operations,"[55] nor does he proffer any explanation for why the D.C. Circuit wouldn't approach such detentions with equal (if not greater) deference to the government's interests—for why the same approach he celebrates in one part of his essay doesn't suggest that judicial review would *not* be disruptive elsewhere. Instead, it's enough merely to assert that "*Maqaleh* requires the judiciary to exercise some humility and defer to most military detention decisions in active theaters of war."[56]

Of course, even this conclusion marginalizes the most relevant fact in *Maqaleh*—that none of the three petitioners were actually seized *in* an "active theater[] of war." Instead, the government *chose* to move the detainees into a theater of active combat operations for the purpose of detention.[57] The D.C. Circuit held that this point was irrelevant to the availability of habeas unless the detainees could prove that the government's *purpose* in so moving them was to avoid judicial review.[58] But logically, if the government's true goal was to avoid judicial interference with active combat operations—rather than to avoid judicial review regardless—moving the detainees *into* an active theater of war seems a rather odd decision, to say the least.

Reasonable minds may well disagree about the result in *Maqaleh*. The larger question that I'm left with after Jacob's response, though, is why we should be so afraid of judicial review. After all, no one has identified a single example in the Guantánamo litigation in which classified information was improperly disclosed by a detainee's counsel. Add that to the fact that the government has prevailed in every case in which it appealed a district court's grant of habeas relief or in which the detainee appealed the denial. Taken together, these points bespeak a record in which judicial review has done exceedingly little to jeopardize the government's interests. Indeed, it may have had the opposite effect, as I described in my initial contribution, of lending legitimacy to our detention program both at Guantánamo and else-

---

55. *Id.*
56. *Id.*
57. *See al-Maqaleh*, 605 F.3d at 87.
58. *See id.* at 98–99.

where. At minimum, it has had the salutary effect of requiring the government to make its case before a neutral magistrate, something that, in the case of an overwhelming majority of the men who since have been released from Guantánamo, it declined to even attempt.

Ultimately, I suspect Jacob and I will have to agree to disagree when it comes to the rationales that the D.C. Circuit has employed in reaching these results, and perhaps even with regard to some of the results themselves. But at least based on the existing record, it strikes me as little more than a canard to conclude that we should accept "the likelihood . . . of erroneous detentions" outside Guantánamo as a necessary evil because judicial review endangers our safety and security. If there is one lesson of which we should take particular heed for detainee policy going forward, it is that judicial review has done no such thing.

Detention Policies: Reply to Stephen Vladeck

Greg Jacob

Professor Vladeck simply cannot imagine how judicial review of military detentions, even in active theaters of war, could possibly disrupt the government's war efforts. If the courts order that detainees be released, then judicial review was clearly necessary to correct erroneous detentions. And if the courts do not order that detainees be released, then what does the government have to complain about? By this standard, judicial review of military detentions is always justified, without regard to cost or outcome.

But of course, this standard does not measure the true cost of judicial review. It must be remembered that the kind of judicial review at issue here was not carefully constructed and balanced by our political leaders, but rather was imposed by the courts as a matter of *constitutional requirement*. The War on Terror and the wars in Afghanistan and Iraq are wars of choice waged against vastly outmatched opponents, but *constitutional requirements* apply equally during wars of necessity in which the nation's very survival is at stake. We held hundreds of thousands of prisoners of war during the Civil War and

World War II—how is Professor Vladeck's expansive judicial review supposed to be administered under such circumstances without seriously compromising our security interests? No practicable answer is even remotely suggested in my sparring partner's essay.

Until the new kind of war presented by the War on Terror came along, the courts uniformly recognized that war is a matter best handled by the political branches, and that at least in active theaters of combat operations, the judiciary should stay out.[59] That is why the D.C. Circuit's decision in *al-Maqaleh* is so important: It recognizes there are times and places in which the substantial costs in time, energy, and resources that necessarily accompany the judiciary's error-correcting function simply aren't worth it, and to which the Framers accordingly never intended to extend constitutional habeas protections. To be sure, the circumstances in which constitutional habeas protections do not apply are carefully circumscribed; U.S. citizens, for example, will always be entitled to habeas review, and after *Boumediene,* most if not all aliens detained domestically will be as well. But within that narrow sphere from which the judiciary has been excluded, and has by and large accepted its exclusion, the time, energy, and resources at stake can literally be a matter of life or death for our troops, and for the nation as a whole.

Professor Vladeck does not believe the stakes could possibly be so high. And in a war of choice in which only a few hundred detainees being held an ocean away from the front lines are permitted access to our courts, perhaps they are not.[60] But how could the military possibly have defended hundreds of thousands of habeas petitions in the midst of World War II? With the witnesses to captures primarily being front-line soldiers still engaged in fighting, should we pull half of Easy Company out of Operation Market Garden to type up detention affidavits? With military intelligence attempting to

---

59. *See, e.g.,* Johnson v. Eisentrager, 339 U.S. at 763, 779 (1950).

60. While we can certainly afford the monetary cost of the Guantanamo habeas litigation, it is not at all clear that the substantial disruption to intelligence-gathering caused by the constant intervention of lawyers and ongoing discovery might not at some point cost us timely access to information needed to save lives.

secure mission-critical information that could turn the tide of war, should we allow their operations to be chilled and disrupted by a stream of discovery requests? And having disarmed enemy troops on the battlefield and placed them in detention camps where they can do no further harm, should we re-arm them with legal causes of action that will consume significant time and manpower to defend, and further provide them a public platform from which to denounce the United States? In light of these costs and disruptions, it is unsurprising that the Geneva Conventions, for example, do not even hint at any kind of judicial review requirement for the ordinary detention of military prisoners.

Will mistakes be made in war, including erroneous detentions? Certainly. Would robust judicial review correct those errors? Some of them, probably. Courts are no more infallible than military review boards, however, and the fact that the D.C. Circuit has overturned every award of habeas relief the government has appealed shows that courts get it wrong plenty of the time, too—having reached opposite conclusions in those cases, the district court and the court of appeals cannot both be right. In the end, however, Professor Vladeck simply presents no evidence that the D.C. Circuit's habeas review procedures have failed to provide adequate error correction for the Guantánamo detainees, or that the expected benefits of extending such review to active theaters of war would outweigh the attendant costs.

Contributors

Norman Abrams served as Acting Chancellor of UCLA from 2006 to 2007. He retired from the faculty of the UCLA Law School in 2007, with the titles of Acting Chancellor Emeritus and Distinguished Professor of Law Emeritus, but he continues to both teach and write in the areas of antiterrorism law, federal criminal law, and evidence. He received his AB and JD degrees from the University of Chicago in 1952 and 1955, respectively, where he served as editor-in-chief of the *University of Chicago Law Review*. He was appointed to the UCLA law school faculty in 1959. Before coming to UCLA, he served as an associate in law at Columbia University Law School and as a research associate and Director of the Harvard-Brandeis Cooperative Research for Israel's Legal Development at the Harvard Law School. While on leave from UCLA in 1966–1967, he served as a Special Assistant to the Attorney General of the United States in the Criminal Division of the U.S. Department of Justice. He also served as Interim Dean of the UCLA Law School in 2003–2004. The fifth edition of Abrams' groundbreaking casebook on federal criminal law, *Federal Criminal Law and Its Enforcement* (with Beale and Klein), was published in 2010. Another of his books, *Antiterrorism and Criminal Enforcement* (4th Ed., 2012), is the first casebook to deal comprehensively with the rapidly evolving field of antiterrorism law and the criminal enforcement process. He is also a co-author of *Evidence—Cases and Materials*, 9th Ed. (with Weinstein, Mansfield, and Berger).

Stewart Baker is a partner in the law firm of Steptoe & Johnson. He is the author of *Skating on Stilts—Why We Aren't Stopping Tomorrow's Terrorism,* a book on the security challenges posed by technology and the use of data in preventing terrorism. From 2005 to 2009, he was the first Assistant Secretary for Policy at the Department of Homeland Security. Baker's practice covers cybersecurity, national security, CFIUS, electronic surveillance, law enforcement, export control, encryption, and related technology issues. From 1992 to 1994, Baker was General Counsel of the National Security Agency, where he led NSA and interagency efforts to reform commercial encryption and computer security law and policy.

Valerie Caproni is currently Vice President and Deputy General Counsel of Northrop Grumman Corporation. Caproni began her legal career as a law clerk for the Honorable Phyllis Kravitch, U.S. Court of Appeals for the Eleventh Circuit. After several years as a litigator at Cravath, Swaine & Moore, she joined the U.S. Attorney's Office, Eastern District of New York, as a federal prosecutor. Although she left the U.S. Attorney for several years to be General Counsel of the New York State Urban Development Corp. (now the New York State Empire Development Corp.), she eventually returned to the U.S. Attorney's Office, where she served in a variety of supervisory positions, including Chief of the Criminal Division. In the U.S. Attorney's Office, Caproni supervised and prosecuted a wide variety of cases, including traditional organized crime, narcotics, white collar, and civil rights. In 1998, Caproni left the U.S. Attorney's Office and joined the Securities and Exchange Commission as Regional Director for the Pacific Regional Office. After several years of managing the regulatory and enforcement activities of the SEC on the West Coast, she returned to private practice. Two years later, she was tapped by Robert Mueller to be General Counsel of the Federal Bureau of Investigation. As the FBI's General Counsel, Caproni provided substantial legal advice and guidance during the FBI's post–9/11 transformation and was involved in a number of legislative and policy initiatives designed to ensure that the FBI was able to achieve its mission as a domestic intelligence agency.

James X. Dempsey is the Vice President for Public Policy at the Center for Democracy and Technology (CDT). From 2003 to 2005, he served as Executive Director; he currently heads CDT West in San Francisco. At CDT, Dempsey concentrates on Internet privacy, government surveillance, and national security issues. Prior to joining CDT, Dempsey was Deputy Director of the nonprofit Center for

Contributors

National Security Studies and Special Counsel to the National Security Archive, a non-governmental organization that uses the Freedom of Information Act to gain the declassification of documents on U.S. foreign policy. From 1985 to 1995, Dempsey was Assistant Counsel to the House Judiciary Committee's Subcommittee on Civil and Constitutional Rights. He worked on issues at the intersection of national security and constitutional rights, including terrorism, counterintelligence, oversight of the Federal Bureau of Investigation, and electronic surveillance laws, as well as criminal justice issues. Dempsey has been a member of the Markle Foundation Task Force on National Security in the Information Age (2004 to the present); the Bill of Rights Defense Committee Advisory Board (2002 to the present); the Board of Directors of the Defending Dissent Foundation (2007 to the present); the Industry Advisory Board for the National Counter-Terrorism Center (2005–2006); and the Transportation Security Administration's Secure Flight Working Group (2005).

Charles J. Dunlap Jr., the former Deputy Judge Advocate General of the U.S. Air Force, joined the Duke Law faculty in July of 2010. Dunlap retired from the Air Force in June of 2010, having attained the rank of major general during a 34-year career in the Judge Advocate Corps. In his capacity as Deputy Judge Advocate General from May of 2006 to March of 2010, he assisted the Judge Advocate General in the professional supervision of more than 2,200 judge advocates, 350 civilian lawyers, 1,400 enlisted paralegals, and 500 civilians around the world. In addition to overseeing an array of military justice, operational, international, and civil law functions, he provided legal advice to the Air Staff and commanders at all levels. In the course of his career, Dunlap has been involved in various high-profile interagency and policy matters, highlighted by his testimony before the U.S. House of Representatives concerning the Military Commissions Act of 2006. Dunlap's legal scholarship also has been published in the *Stanford Law Review*, the *Yale Journal of International Affairs*, the *Wake Forest Law Review*, the *Fletcher Forum of World Affairs*, the *University of Nebraska Law Review*, the *Texas Tech Law Review*, and the *Tennessee Law Review*, among others.

Louis Fisher is Scholar in Residence at the Constitution Project. Previously, he worked for four decades at the Library of Congress as Senior Specialist in Separation of Powers (Congressional Research Service, from 1970 to 2006) and Specialist in Constitutional Law (from 2006 to 2010). During his service with CRS, he was Research Director of the House Iran-Contra Committee in 1987, writing major sections of the final report. He has twice won the Louis Brownlow Book Award; the encyclopedia he co-edited was awarded the Dartmouth Medal. In 1995 he received the Aaron B. Wildavsky Award For Lifetime Scholarly Achievement in Public Budgeting from the Association for Budgeting and Financial Management; and in 2006 he received the Neustadt Book Award for *Military Tribunals and Presidential Power: American Revolution to the War on Terrorism* (2005). Dr. Fisher has been invited to testify before Congress on such issues as war powers, the state secrets privilege, NSA surveillance, executive spending discretion, presidential reorganization authority, Congress and the Constitution, the legislative veto, the item veto, the Gramm-Rudman-Hollings Act, executive privilege, committee subpoenas, executive lobbying, CIA whistleblowing, covert spending, the pocket veto, recess appointments, the budget process, the balanced budget amendment, biennial budgeting, and presidential impoundment powers.

Michael German serves as Policy Counsel for National Security, ACLU Washington Legislative Office. He was a Special Agent for the Federal Bureau of Investigation from 1988 to 2004. He has taught counterterrorism at the FBI National Academy and worked as an undercover agent against white supremacist groups in Los Angeles, California, and militia groups in Seattle, Washington. He resigned from the FBI in June after protesting the handling of a terrorism case. German is the author of *Thinking Like a Terrorist: Insights of a Former FBI Undercover Agent* (Potomac Books, 2007), which provides unique insights into why terrorism is such a persistent and difficult problem and why the U.S. approach to counterterrorism is not working. His articles on terrorism have appeared in the *Washington Post,* the *San Francisco Chronicle,* and the *National Law Journal.* For his work combating white supremacist groups, he was awarded the Los Angeles Federal Bar Association Medal of Valor and the First African Methodist Episcopal Church FAME Award.

Amos N. Guiora is a professor of law at the S.J. Quinney College of Law, University of Utah. Guiora, who teaches criminal procedure, international law, global perspectives on counterterrorism, and religion and terrorism, incorporates innovative, scenario-based instruction to address national and international security issues and dilemmas. He is a Research Fellow at the International Institute on Counter-Terrorism, The Interdisciplinary Center, Herzliya, Israel; a Corresponding Member, The Netherlands School of Human Rights Research, University of Utrecht School of Law; and was awarded a Senior Specialist Fulbright Fellowship for The Netherlands in 2008. Guiora has published extensively in both the United States and Europe on issues related to national security, limits of interrogation, religion and terrorism, the limits of power, and multiculturalism and human rights. He is the author *of Global Perspectives on Counterterrorism, Fundamentals of Counterterrorism, Constitutional Limits on Coercive Interrogation, Homeland Security: What Is It and Where Are We Going?*, and *Freedom from Religion: Rights and National Security*. He served for 19 years in the Israel Defense Forces as Lieutenant Colonel (retired) and held a number of senior command positions, including Commander of the IDF School of Military Law and Legal Advisor to the Gaza Strip. Guiora was awarded the S.J. Quinney College of Law Faculty Scholarship Award in 2011.

Monica Hakimi is an assistant professor of law at the University of Michigan Law School. Her articles appear in the *Michigan Law Review,* the *American Journal of International Law,* the *European Journal of International Law*, and the *Yale Journal of International Law*. Hakimi earned her JD from the Yale Law School in 2001 and her BA, summa cum laude, from Duke University. Following law school, she clerked for Judge Kimba Wood of the Southern District of New York and later served as an attorney-adviser in the Office of the Legal Adviser at the U.S. Department of State.

Bernard Horowitz, a 2010 recipient of Colorado College's Hohback Award for Distinction in Political Science, serves as a Research Fellow with the ABA Standing Committee on Law and National Security. With Harvey Rishikof, he has co-written "Reflections on Terrorism" for the *University of Minnesota Law Review* and "Enemy Combatant Detainees," a section of *Congress and the Politics of National Security* (edited by David Auerswald and Colton Campbell, Cambridge University Press, 2012).

Greg Jacob is a partner with O'Melveny & Myers LLP. Jacob served in several high-profile positions in the federal government, including at the White House, Department of Justice, and Department of Labor. Most recently, Jacob served as Solicitor of Labor, the third-ranking official in the Department of Labor. In this position, Jacob oversaw more than 600 employees, including 425 attorneys, and broadly managed the department's most important litigation under more than 180 federal labor and employment laws. Before joining the Department of Labor, Jacob served as Senior Advisor to the Secretary of Labor, Deputy Solicitor of Labor, Attorney Advisor in the Department of Justice's Office of Legal Counsel, and as Special Assistant to President George W. Bush, with responsibility for formulating policy on immigration, justice, disability, tort reform, and other legal issues.

Orin Kerr serves as professor of law at George Washington Law School. He teaches criminal law, criminal procedure, and computer crime law. His articles have appeared in the *Harvard Law Review, Yale Law Journal, Stanford Law Review, Columbia Law Review, University of Chicago Law Review, Michigan Law Review, Virginia Law Review, New York University Law Review, Georgetown Law Journal, Northwestern University Law Review, Texas Law Review*, and many other journals. According to the most recent *Leiter Rankings,* Kerr is ranked number seven among criminal law scholars in the United States for citations in academic journals. His scholarly articles have been cited by all the regional U.S. courts of appeals and many federal district courts. Before joining the GW Law faculty in 2001, Kerr was an honors program trial attorney in the Computer

Crime and Intellectual Property Section of the Criminal Division at the U.S. Department of Justice, as well as a Special Assistant U.S. Attorney for the Eastern District of Virginia. He is a former law clerk for Justice Anthony M. Kennedy of the U.S. Supreme Court and Judge Leonard I. Garth of the U.S. Court of Appeals for the Third Circuit. In the summer of 2009 and 2010, he served as Special Counsel for Supreme Court nominations to Senator John Cornyn on the Senate Judiciary Committee. The GW Law Class of 2009 awarded Kerr the Distinguished Faculty Service Award, the law school's teaching award.

Susan Landau works in the areas of cybersecurity, privacy, and public policy. Landau was a Distinguished Engineer at Sun Microsystems and has been a faculty member at the University of Massachusetts at Amherst and at Wesleyan University. She has held visiting positions at Harvard, Cornell, and Yale, and the Mathematical Sciences Research Institute. Landau is the author of *Surveillance or Security? The Risks Posed by New Wiretapping Technologies* (MIT Press, 2011) and coauthor, with Whitfield Diffie, of *Privacy on the Line: The Politics of Wiretapping and Encryption* (MIT Press, 1998, Rev. Ed., 2007), and has written numerous computer science and public policy papers, as well as op-eds on cybersecurity and encryption policy. In 2011, Landau testified for the House Judiciary Committee on security risks in wiretapping, while in 2009 she testified for the House Science Committee on Cybersecurity Activities at NIST's Information Technology Laboratory. Landau serves on the Computer Science and Telecommunications Board of the National Research Council and on the advisory committee for the National Science Foundation's Directorate for Computer and Information Science and Engineering. She also served on the Commission on Cyber Security for the 44th Presidency and on the Information Security and Privacy Advisory Board. Landau was a fellow at the Radcliffe Institute for Advanced Study, is the recipient of the 2008 Women of Vision Social Impact Award, and a fellow of both the American Association for the Advancement of Science and the Association for Computing Machinery.

Gordon Lederman is currently a congressional staffer. From 1998 to 2003, Lederman was a member of the National Security Law and Policy Practice Group of Arnold & Porter in Washington, DC. He advised clients on the nexus of national security, law enforcement, technology, civil liberties, and privacy, and practiced in international law. In 2003, Lederman joined the 9/11 Commission staff and was responsible for assessing the Intelligence Community's senior-level management structure. After the 9/11 Commission released its report in July of 2004, Lederman moved to the Senate Homeland Security and Governmental Affairs Committee as a Special Bipartisan staffer. In that capacity, he served as a leading Senate staff drafter and negotiator of the Intelligence Reform and Terrorism Prevention Act of 2004, which enacted the Commission's recommendations to create the Director of National Intelligence and National Counterterrorism Center. He also worked on the Senate's investigation of governmental preparedness for and response to Hurricane Katrina. He subsequently joined the National Counterterrorism Center to help implement that legislation. He then practiced international law at White & Case in Washington, D.C., served as the Director of Legal Affairs for the Project on National Security Reform in Washington, D.C., and returned to Congress. Lederman has published widely in the area of national security policy and organizational reform, including *Reorganizing the Joint Chiefs of Staff: The Goldwater-Nichols Department of Defense Reorganization Act of 1986*.

Kate Martin has been Director of the Center for National Security Studies, a nonprofit human rights and civil liberties organization located in Washington, D.C., since 1992. Previously, she was a litigation partner in the law firm of Nussbaum, Owen and Webster. From 1993 to 1999, Martin was also co-director of a project on Security Services in a Constitutional Democracy in 12 former communist countries in Europe. Martin has taught strategic intelligence and public policy at Georgetown University Law School and national security law at George Washington School of Law. She also served as General Counsel to the National Security Archive, a research library located at George Washington University from 1995 to 2001. Since 1988, she has testified and written extensively on the intersections of national security and civil liberties.

Greg Nojeim is a Senior Counsel at the Center for Democracy & Technology (CDT) and the Director of its Project on Freedom, Security & Technology. Nojeim conducts much of CDT's work in the areas of national security, terrorism, and Fourth Amendment protections. He is deeply involved in a multiyear effort to bring Fourth Amendment privacy protections to digital communications. He frequently testifies before Congress on civil liberties issues connected to cybersecurity and antiterrorism legislation such as the PATRIOT Act. Nojeim is former Co-chair of the Coordinating Committee on National Security and Civil Liberties of the Individual Rights and Responsibilities Section of the American Bar Association. He currently sits on the DHS Data Privacy and Integrity Advisory Committee. Prior to joining CDT in May of 2007, Nojeim worked for 12 years in the Washington Legislative Office of the American Civil Liberties Union as a Legislative Counsel and as its Associate Director and Chief Legislative Counsel. Nojeim was also employed for five years as an attorney with the Washington, D.C., law firm of Kirkpatrick & Lockhart (now Kirkpatrick & Lockhart Preston Gates Ellis LLP), where he specialized in mergers and acquisitions, securities law, and international trade.

Michelle Richardson is a Legislative Counsel with the Washington Legislative Office of the American Civil Liberties Union, focusing on national security and government transparency. She monitors and analyzes legislation and executive branch policy concerning the PATRIOT Act, the Foreign Intelligence Surveillance Act, cybersecurity, state secrets, whistleblower protection, and the Freedom of Information Act. Richardson is responsible for drafting legislative proposals, and crafting the ACLU's congressional testimony and other formal communications to Congress and the administration, and representing and advocating for the ACLU's position in regular meetings and communications with members of Congress and the administration and their staffs. Richardson also plays a key role in the design of lobbying, grassroots, and messaging strategies for ACLU members and activists and for legislative coalitions. She has provided legislative and political analysis and commentary to media outlets, blogs, and trade publications such as the AP, UPI, *Washington Post, C-SPAN, CBS News, Time, CQ, Politico, Al Jazeera, NPR, Wired, Democracy Now, Talking Points Memo,* and the *Huffington Post.* Before coming to the ACLU in 2006, Richardson served for three years as Counsel to the House Judiciary Committee, where she specialized in national security, civil rights, and constitutional issues for Democratic ranking member John Conyers. Her work included drafting legislation and committee reports and conducting oversight of the Department of Justice's post–9/11 antiterrorism policies. Richardson is a graduate of the University of Colorado at Boulder and American University's Washington College of Law.

Harvey Rishikof is Chair of the American Bar Association Standing Committee on Law and National Security and was a professor of law and national security studies at the National Defense University, National War College in Washington, D.C., where he chaired the Department of National Security Strategy. He is a lifetime member of the American Law Institute and the Council on Foreign Relations. Rishikof was a federal law clerk in the Third Circuit for the Honorable Leonard I. Garth, a social studies tutor at Harvard University, attorney at Hale and Dorr, Administrative Assistant to the Chief Justice of the United States, legal counsel for the Deputy Director of the Federal Bureau of Investigation, and Dean of Roger Williams School of Law. Currently, he is also an adviser to the *Harvard Law Journal on National Security* and serves on the Board of Visitors at the National Intelligence University. He has written numerous articles, law reviews, and book chapters. He and Roger George recently co-authored *The National Security Enterprise: Navigating the Labyrinth* (Georgetown Press, 2011).

Paul Rosenzweig is the founder of Red Branch Consulting PLLC, a homeland security consulting company, and a Senior Advisor to The Chertoff Group. Rosenzweig formerly served as Deputy Assistant Secretary for Policy in the Department of Homeland Security. He is a Distinguished Visiting Fellow at the Homeland Security Studies and

Analysis Institute. He also serves as a professorial lecturer in law at George Washington University, a senior editor of the *Journal of National Security Law & Policy,* and as a Visiting Fellow at The Heritage Foundation. In 2011, he was a Carnegie Fellow in National Security Journalism at the Medill School of Journalism at Northwestern University. Rosenzweig is a cum laude graduate of the University of Chicago Law School. He has an MS in Chemical Oceanography from the Scripps Institution of Oceanography, University of California at San Diego, and a BA from Haverford College. Following graduation from law school, he served as a law clerk to the Honorable R. Lanier Anderson III of the U.S. Court of Appeals for the Eleventh Circuit. He is the co-author (with James Jay Carafano) of *Winning the Long War: Lessons from the Cold War for Defeating Terrorism and Preserving Freedom* and author of the forthcoming *Cyber Warfare: How Conflicts in Cyberspace Are Challenging America and Changing the World.*

Anthony (Tony) Rutkowski is CEO at Netmagic Associates LLC and Executive Vice President for Industry Standards and Regulatory Affairs at Yaana Technologies, a Silicon Valley security and forensics company. He currently leads the cybersecurity standards activity at the Geneva-based ITU-T, heads the ETSI electronic warrant standards project, and has written an array of key international standards for cybersecurity, network forensics, and identity management capabilities. Over the past 10 years, he has participated extensively in most of the principal international, EU, and domestic U.S. industry, standards, and regulatory activities in the lawful interception and retained data fields. He also lectures occasionally at the Georgia Tech Nunn School as a Distinguished Senior Research Fellow. His career as an engineer-lawyer spans 45 years in many private-sector, government, and academic positions in the United States and abroad.

Steven Siegel was recently appointed Senior Counsel for Investigations in the General Counsel's office at Northrop Grumman. For eight years, until April of 2012, he served in the FBI's Office of the General Counsel where, since September of 2009, he was the Deputy General Counsel for the National Security Law Branch. While at the

FBI, he also held the positions of Section Chief for Policy, Litigation, Oversight and Compliance and Chief of the Classified Litigation Support Unit, and he also served as a line attorney in the Counterintelligence Law Unit and a Counterterrorism Law Unit. Prior to joining the FBI, Siegel was a prosecutor for approximately 11 years. He began his career as an Assistant District Attorney in Queens County, New York, where his final assignment was prosecuting organized-crime cases. Siegel also served as a federal prosecutor in the U.S. Attorney's Office for the Eastern District of New York, where he primarily handled narcotics cases. Immediately prior to joining the FBI, Siegel was a trial attorney in the Department of Justice, Criminal Division, Narcotic and Dangerous Drug Section, where he prosecuted international narcotics cases.

Christopher Slobogin occupies the Milton Underwood Chair at Vanderbilt Law School, where he directs the Criminal Justice Program. He has authored scores of articles on criminal procedure and has written both a treatise and a textbook on the subject; according to the *Leiter Report,* he is one of the 10 most cited criminal law and procedure law professors in the nation. Before joining Vanderbilt's law faculty in 2008, Slobogin held the Stephen C. O'Connell Chair at the University of Florida's Levin College of Law; he also taught at Stanford, the University of Southern California, and the University of Virginia Law Schools. He has appeared on *Good Morning America, Nightline, the Today Show, National Public Radio,* and many other media outlets, and has been cited in more than 2,000 law review articles or treatises and more than 100 judicial opinions, including opinions at the Supreme Court level. Slobogin holds a secondary appointment as a professor in the Vanderbilt School of Medicine's Department of Psychiatry.

Stephen I. Vladeck is a professor of law and the Associate Dean for Scholarship at American University Washington College of Law. A nationally recognized expert on the role of the federal courts in the war on terrorism, he was part of the legal team that successfully challenged the Bush administration's use of military tribunals at Guantánamo Bay, Cuba, in *Hamdan v. Rumsfeld,* and has co-authored amicus briefs in a host of other major lawsuits, many of which have challenged the U.S. government's surveillance and detention of terrorism suspects. Vladeck, who is a co-editor of Aspen Publishers' leading national security law casebook, has drafted reports on related topics for a wide range of organizations, including the First Amendment Center, the Constitution Project, and the ABA's Standing Committee on Law and National Security. Vladeck has won awards for his teaching, his scholarship, and his service to the law school. He is a senior editor of the peer-reviewed *Journal of National Security Law and Policy,* a senior contributor to the *Lawfare* blog, and a member of the Executive Committee of the Section on Federal Courts of the Association of American Law Schools. Vladeck clerked for the Honorable Marsha S. Berzon on the U.S. Court of Appeals for the Ninth Circuit and the Honorable Rosemary Barkett on the U.S. Court of Appeals for the Eleventh Circuit. While a law student, he was executive editor of the *Yale Law Journal.*

John Yoo is a professor of law at the University of California Berkeley School of Law and a Visiting Scholar at the American Enterprise Institute. After clerking for Justice Clarence Thomas of the U.S. Supreme Court, he served as General Counsel of the U.S. Senate Judiciary Committee from 1995 to 1996. From 2001 to 2003, he served as a Deputy Assistant Attorney General in the Office of Legal Counsel at the U.S. Department of Justice, where he worked on issues involving foreign affairs, national security, and the separation of powers. Yoo has received the Paul M. Bator Award for excellence in legal scholarship and teaching from the Federalist Society for Law and Public Policy. He is the author of *The Powers of War and Peace: The Constitution and Foreign Affairs after 9/11* (University of Chicago Press, 2005); *War by Other Means: An Insider's Account of the War on Terror* (Grove/Atlantic, 2006); *Crisis and Command: The History of Executive Power from George Washington to George W. Bush* (Kaplan, 2010); *Confronting Terror: 9/11 and the Future of American National Security* (Encounter, 2011) (with Dean Reuter); and *Taming Globalization: International Law, the U.S. Constitution, and the New World Order* (Oxford University Press, 2012).